THE HISTORY *of* MEDICINE

EARLY CIVILIZATIONS

PREHISTORIC TIMES TO 500 C.E.

THE HISTORY of MEDICINE

EARLY CIVILIZATIONS

PREHISTORIC TIMES TO 500 C.E.

KATE KELLY

EARLY CIVILIZATIONS: Prehistoric Times to 500 C.E.

Facts On File, Inc.
An imprint of Infobase Publishing
132 West 31st Street
New York NY 10001

Library of Congress Cataloging-in-Publication Data
Kelly, Kate, 1950–
Early civilizations : prehistoric times to 500 C.E. / Kate Kelly.
p. cm. — (The history of medicine)
Includes bibliographical references and index.
ISBN 978-0-8160-7205-7
1. Medicine, Ancient. I. Title.
R135.K45 2009
610.938—dc22 2008043441

Facts On File books are available at special discounts when purchased in bulk quantities for businesses, associations, institutions, or sales promotions. Please call our Special Sales Department in New York at (212) 967-8800 or (800) 322-8755.

You can find Facts On File on the World Wide Web at http://www.factsonfile.com

Text design by Annie O'Donnell
Illustrations by Bobbi McCutcheon
Photo research by Elizabeth H. Oakes

Printed in the United States of America

Bang FOF 10 9 8 7 6 5 4 3 2 1

This book is printed on acid-free paper.

CONTENTS

PREFACE

"You have to know the past to understand the present."
—*American scientist Carl Sagan (1934–96)*

The history of medicine offers a fascinating lens through which to view humankind. Maintaining good health, overcoming disease, and caring for wounds and broken bones was as important to primitive people as it is to us today, and every civilization participated in efforts to keep its population healthy. As scientists continue to study the past, they are finding more and more information about how early civilizations coped with health problems, and they are gaining greater understanding of how health practitioners in earlier times made their discoveries. This information contributes to our understanding today of the science of medicine and healing.

In many ways, medicine is a very young science. Until the mid-19th century, no one knew of the existence of germs, so as a result, any solutions that healers might have tried could not address the root cause of many illnesses. Yet for several thousand years, medicine has been practiced, often quite successfully. While progress in any field is never linear (very early, nothing was written down; later, it may have been written down, but there was little intra-community communication), readers will see that some civilizations made great advances in certain health-related areas only to see the knowledge forgotten or ignored after the civilization faded. Two early examples of this are Hippocrates' patient-centered healing philosophy and the amazing contributions of the Romans to public health through water-delivery and waste-removal systems. This knowledge was lost and had to be regained later.

The six-volumes in the History of Medicine set are written to stand alone, but combined, the set presents the entire sweep of the history of medicine. It is written to put into perspective for high school students and the general public how and when various medical discoveries were made and how that information affected health care of the time period. The series starts with primitive humans and concludes with a final volume that presents readers with the very vital information they will need as they must answer society's questions of the future about everything from understanding one's personal risk of certain diseases to the ethics of organ transplants and the increasingly complex questions about preservation of life.

Each volume is interdisciplinary, blending discussions of the history, biology, chemistry, medicine and economic issues and public policy that are associated with each topic. *Early Civilizations,* the first volume, presents new research about very old cultures because modern technology has yielded new information on the study of ancient civilizations. The healing practices of primitive humans and of the ancient civilizations in India and China are outlined, and this volume describes the many contributions of the Greeks and Romans, including Hippocrates' patient-centric approach to illness and how the Romans improved public health.

The Middle Ages addresses the religious influence on the practice of medicine and the eventual growth of universities that provided a medical education. During the Middle Ages, sanitation became a major issue, and necessity eventually drove improvements to public health. Women also made contributions to the medical field during this time. The Middle Ages describes the manner in which medieval society coped with the Black Death (bubonic plague) and leprosy, as illustrative of the medical thinking of this era. The volume concludes with information on the golden age of Islamic medicine, during which considerable medical progress was made.

The Scientific Revolution and Medicine describes how disease flourished because of an increase in population, and the book describes the numerous discoveries that were an important aspect of this time. The volume explains the progress made by Andreas Vesalius (1514–64) who transformed Western concepts of the structure of the human body; William Harvey (1578–1657), who studied and wrote about the circulation of the human blood; and Ambroise Paré (1510–90), who was a leader in surgery. Syphilis was a major scourge of this time, and the way that society coped with what seemed to be a new illness is explained. Not all beliefs of this time were progressive, and the occult sciences of astrology and alchemy were an important influence in medicine, despite scientific advances.

Old World and New describes what was happening in the colonies as America was being settled and examines the illnesses that beset them and the way in which they were treated. However, before leaving the Old World, there are several important figures who will be introduced: Thomas Sydenham (1624–89) who was known as the English Hippocrates, Herman Boerhaave (1668–1738) who revitalized the teaching of clinical medicine, and Johann Peter Frank (1745–1821) who was an early proponent of the public health movement.

Medicine Becomes a Science begins during the era in which scientists discovered that bacteria was the cause of illness. Until 150 years ago, scientists had no idea why people became ill. This volume describes the evolution of "germ theory" and describes advances that followed quickly after bacteria was identified, including vaccinations, antibiotics, and an understanding of the importance of cleanliness. Evidence-based medicine is introduced as are medical discoveries from the battlefield.

Medicine Today examines the current state of medicine and reflects how DNA, genetic testing, nanotechnology, and stem cell research all hold the promise of enormous developments within the course of the next few years. It provides a framework for teachers and students to understand better the news stories that are sure to be written on these various topics: What are stem cells, and why is investigating them so important to scientists? And what is nanotechnology? Should genetic testing be permitted? Each of the issues discussed are placed in context of the ethical issues surrounding it.

Each volume within the History of Medicine set includes an index, a chronology of notable events, a glossary of significant terms and concepts, a helpful list of Internet resources, and an array of historical and current print sources for further research. Photographs, tables, and line art accompany the text.

I am a science and medical writer with the good fortune to be assigned this series. For a number of years I have written books in collaboration with physicians who wanted to share their medical knowledge with laypeople, and this has provided an excellent background in understanding the science and medicine of good health. In addition, I am a frequent guest at middle and high schools and at public libraries addressing audiences on the history of U.S. presidential election days, and this regular experience with students keeps me fresh when it comes to understanding how best to convey information to these audiences.

What is happening in the world of medicine and health technology today may affect the career choices of many, and it will affect the health care of all, so the topics are of vital importance. In addition, the public health policies under consideration (what medicines to develop, whether to permit stem cell research, what health records to put online, and how and when to use what types of technology, etc.) will have a big impact on all people in the future. These subjects are in the news daily, and students who can turn to authoritative science volumes on the topic will be better prepared to understand the story behind the news.

ACKNOWLEDGMENTS

This book as well as the others in the set was made possible because of the guidance, inspiration, and advice offered by many generous individuals who have helped me understand better both science and medicine and their histories. I would like to express my heartfelt appreciation to Frank Darmstadt, whose vision and enthusiastic encouragement, patience, and support helped shape the series and saw it through to completion. Thank you, too, to all of the staff members who worked on these volumes.

All of the art and the photography for this set were provided by two very helpful professionals: Bobbi McCutcheon provided all the line art and frequently reached out to me from her office in Juneau, Alaska, to offer very welcome advice and support as we worked through the complexities of the renderings, and Elizabeth Oakes found a wealth of wonderful photographs that helped bring the information to life. Carol Sailors started me off greatly, and Carole Johnson kept me sane by providing able help on the back matter of all the books. Agent Bob Diforio has remained steadfast in his shepherding of the work.

I also want to acknowledge the wonderful archive collections that have provided information for the book. Without places like the Sophia Smith Collection at the Smith College Library, firsthand accounts of the Civil War battlefield treatment or reports such as Lillian Gilbreth's on helping the disabled after World War I would be lost to history.

INTRODUCTION

Natural forces within us are the true healers of disease.

—*Hippocrates*

The story of early medicine is one of magic and sorcery, religion and prayers, shamans and surgeons, ingenuity and experimentation. All manner of successes and failures also dot the backdrop of early medicine. The health challenges of the time were many, and they ranged from near-fatal accidents to a wide variety of mysterious illnesses. Despite very little understanding of how the body worked or why people became sick, primitive people still devised successful methods to help heal the ill and the injured. *Early Civilization: Prehistoric Times to 500 C.E.* tells of what they learned about medicine.

Because there is no written record from prehistoric times, much about early medicine has to be surmised as speculation. The evidence of how prehistoric people lived, what they ate, the sicknesses from which they suffered, and how they treated these illnesses has been gathered from small clues found in various parts of the world beginning as far back as 10,000 years ago. There were no archivists or librarians filing records in boxes in date-order, so an overview of the early history of medicine is fragmentary but still quite fascinating.

Early Civilizations focuses on the story of ancient people and their approach to medicine, but the information presented is by no means limited to discoveries made long ago. Today's scientists, archaeologists, and paleopathologists have learned the value of applying the latest technological developments to examine prehistoric finds. Electron microscopes are being used to identify the scrapings from ancient shards of pottery to reveal what people ate, and computed tomography techniques (CT scans) are being employed to examine mummies to comprehend better the illnesses from which the Egyptians suffered. As a result, there is a high level of present-day excitement among professionals as they gain greater understanding of the diet, health issues, illnesses, and injuries of ancient people. When new clues are found or new technology is created that allows scientists to evaluate old clues better, then the theories of how early people lived and what they suffered from are revised.

Despite progress in these fields of study, there are many questions that are still unanswered, and there are many gaps in today's knowledge of this prehistoric time. What can be known for certain is that early people fell and broke bones, were injured in hunting accidents, had difficulty finding food, had problems with childbirth, suffered illnesses, and endured aches and pains for which they sought relief. When a community member became ill, the entire group became involved. Each person contributed to the group as a whole, so all community members were valued. Hunter-gatherers predominated in early cultures, and an injured hunter who could not hunt or a woman who could not help with watching children or foraging for plants to eat created a great loss for the community.

Readers will learn about the illnesses and injuries that befell primitive people and how they searched for cures (chapter 1) during the period of time covered by *Early Civilizations* (approximately 10,000 years ago through 500 C.E., the end of the Roman Empire). Chapter 2 introduces the knowledge gained by the Egyptians, who contributed to the future by transcribing many of their medical procedures as well as by creating a preservation method—mummification—that has permitted Egyptian bodies to last for thousands of years. Egyptians also created elaborate burial sites that have proved illuminating to present-day archaeologists who look to more fully understand matters of health and causes of death that affected ancient people.

Indian medical knowledge is explored in chapter 3, where readers will learn of great advances in medical treatment as well as significant progress in surgical methods, particularly in the arena of plastic surgery. The next chapter, chapter 4, explains how China's isolated position in the ancient world meant that no one paid much attention to its culture and its discoveries until more recently. Despite years of being overlooked, Chinese medicine now wields influence on a number of medical practices of today: Acupuncture and acupressure are just two of its contributions.

Chapter 5 examines how Greek civilization advanced medical treatment. Early influences ranged from an understanding of the importance of a healthy diet and exercise to beliefs such as this one: If a sick person sleeps in a specific temple in a community, then prophetic dreams, interpreted by the attending priest, would lead to a healing cure.

Two very influential Greek medical leaders, Hippocrates and Galen, each merits a chapter of his own. Hippocrates developed a patient-centric curative method that involved observation and encouragement of natural healing. Galen contributed greatly to an understanding of the anatomy and its importance in health. He also popularized the belief in the "four humors" of medicine, which distracted medical practitioners from searching for more substantial medical knowledge. Galen, unfortunately, cast a very long shadow on medical progress and inhibited scientific advancement for thousands of years because the belief in his theories was so deep.

The final chapter of *Early Civilizations* addresses progress made during the Roman Empire. The Romans did very little to increase knowledge of medical treatment, but they furthered civilization in another way that had an effect on the people's health. Because they valued good sanitation, the Romans built cities that featured clean drinking water and methods for proper disposal of garbage and rubbish. As a result, their people were less likely to be exposed to some forms of disease.

Although many of the healing methods used by ancient people may appear primitive today, it is important to remember that it was to be thousands of years—not until the 19th century—before doctors would have any understanding of the role germs played in disease. When it is considered that this vital piece of information was unknown until relatively recently, it is remarkable that early humanity was able to develop any methods at all that encouraged natural healing.

Early Civilizations: Prehistoric Times to 500 C.E. has been written to illuminate what occurred during the ancient times that affected future developments in medicine. The back matter contains a chronology, a glossary, and an array of historical and current sources for further research. These sections should prove especially helpful for readers who need additional information on specific terms, topics, and developments in medical science.

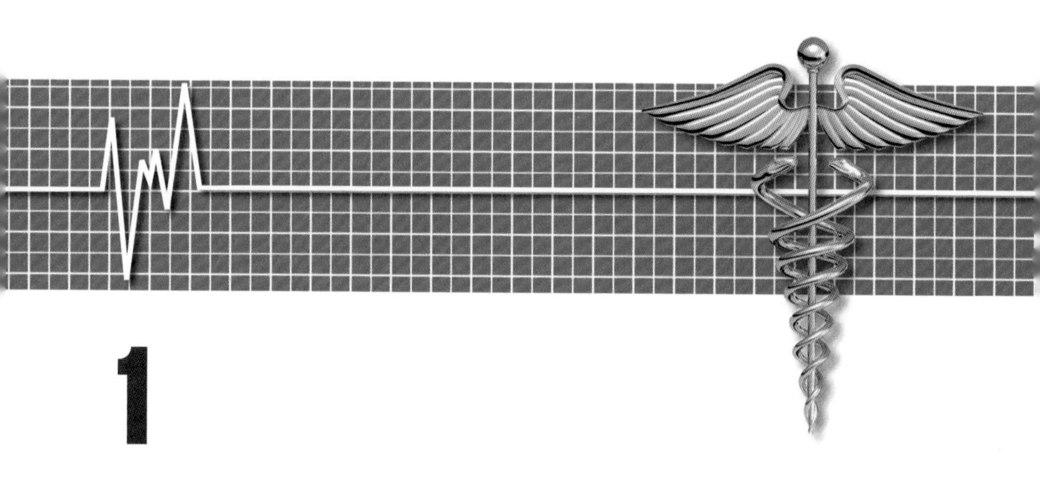

1

Paleopathology: The Study of Disease in Prehistoric Times

Early humans were hunter-gatherers. Tracking wildlife, primitive people eventually traveled to every continent except present-day Antarctica as they hunted herds of wild animals and sought out plants to eat along the way. Living in widely scattered clusters of 50 to 100 people, they relied on each member of the group for overall survival. From the beginning of this early existence, humans needed to maintain good health and to take adequate care of injuries and illnesses to survive. This was not an easy task as their lives were dangerous. When a group member became ill or was hurt, the others had to devise healing methods through trial and error.

While the specific geographic challenges varied from group to group—some clusters of people might have faced draught or beastly heat while others might have had to maintain an existence in an area with ice and snow—the basic trials that these different people faced were very similar. They all needed food, shelter, and the ability to care for the very young as well as for the older members of their group. (The life spans of people during this time would have been quite short.) During difficult times, hunger may have completely wiped out some populations.

There is no written record of the earliest prehistoric people, their illnesses, or their medical treatments; however, information that was patched together from this time by scientists and historians provides a basic understanding of prehistoric people's health and lifestyles. Many of the healing methods that were developed by people in all parts of the world were more similar than might be expected, considering the broad and diverse geographic area that was inhabited by prehistoric people. Magic, religion, and experimentation with plants as healing agents played a prominent role in medical practices everywhere. This chapter begins with a look at how prehistorians and paleopathologists of today learn about the health issues and medical beliefs of early peoples. It will continue with an examination of the early diseases—and the treatments—that affected hunter-gatherers and how health problems changed when people began to farm and to live together in communities like the towns that grew up in Mesopotamia.

THE IMPORTANCE OF UNDERSTANDING EARLY HUMANS AND THEIR HEALTH

The study of ancient people and their illnesses is known as *paleopathology.* Although the focus of this science is on old and "dusty" specimens, the field itself is one that is brimming with new discoveries. Cutting-edge medical techniques are being used on skeletons and remains from long ago, and scientists are delving into the past to comprehend current and emerging diseases of today better.

The knowledge of where diseases—even ancient ones—come from, why and how they spread, and what conditions make them particularly deadly is very useful. In every geographic area, paleopathologists examine the diseases that were present and the significance of a particular disease to those inhabitants. Studying diseases from the past also provides helpful information about the diseases themselves and how they spread. Egyptians today, for example, suffer from a parasite *schistosomiasis,* evidence of which is commonly found in mummies as well. Scientists are using DNA analysis to compare the ancient form of the organism with the one that plagues modern Egyptians to determine whether or not the parasite has evolved over time. New treatments may be based on what is revealed.

Paleopathology also sheds light on history. Populations ebb and flow throughout every era, and an understanding of a people's state of health helps explain what happened to them in the course of time. Were they beaten on a battlefield, or were they decimated by a disease? Scientists and historians are still trying to determine whether the collapse of the Mayan civilization was due to conquest by outsiders, an epidemic that spread through its communities, climate change, or the civilization's inability to feed its citizenry because of soil exhaustion. In his recent book, *Collapse: How Societies Choose to Fail or Succeed,* Jared Diamond notes that the answer may vary somewhat from Mayan site to site but also that multiple factors combined to bring an end to what was a very advanced civilization. What they learn as studies continue may affect the world of medicine or the way people decide to manage the environment.

Evidence for Study

Since soft tissue decomposes in time, bones and teeth are the most commonly available clues as to what early people's lives were like and what some of their health problems were. While skeletons cannot provide information about diseases of the soft tissues or organs, paleopathologists find that bones can tell about a person's state of health. If a person is malnourished, for example, the bone marrow will show a depletion of iron because the body must draw from reserve iron stores to survive. Hereditary disorders and illnesses such as tuberculosis, brucellosis, and leprosy also reveal themselves through skeletal lesions. Scientists prefer to study full skeletons rather than isolated bones; the more complete the skeleton, the more complete the story. Scientists also can draw more accurate information when they can study not just one but several members of a population.

Paleopathologists have benefited greatly from the Egyptian practice of mummification. (This is fully examined in the next chapter, "What the Mummies Reveal.") The studies of these preserved remains are providing fascinating information about these ancient people (this topic will be explored fully in the following chapter). Another source of information has been "peat-bog people." Well-preserved bodies from the Bronze (2000 B.C.E.–800 B.C.E.) and Iron Ages (800 B.C.E.–0 C.E.) have been found in the peat bogs of Germany, Denmark, Ireland, and England. Historians think that ancient people believed that the peat bogs were magical, and they came to these areas to perform sacrifices. As it happens, the damp, cool soil of ancient peat bogs has served as an excellent preservative. (Peat is an accumulation of partially decayed vegetation that forms in wetland areas, which can be cut out, dried, and used as fuel.) Unfortunately, the peat destroys DNA, so while the bones—albeit changed into what is described as a rubbery form—and skin may remain, it has been difficult to deduce other information from peat-bog bodies.

In addition, the amazing discovery of the "iceman" (see the section entitled "The Discovery of the Iceman" later in this chapter) in northern Italy has revealed a great deal that has heretofore been unknown. Frozen in snow and not tampered with since the time of his death in about 3300 B.C.E., this ancient specimen has now undergone scrutiny by scientists wielding the latest technology.

Grauballe-Man, here at the Mosegaard-Museum, Denmark, was discovered in 1952 in Nebelgård Mose, a bog in Jutland, Denmark. Bronze and Iron Age bodies that have been recovered from peat bogs are quite well preserved. *(Malene Thyssen)*

The information gathered has been beneficial to historians and scientists alike.

A few Stone Age cultures still exist in very remote parts of Africa, New Guinea, and Australia. (Modern civilization has spread to the point that these groups of people are increasingly rare.) Anthropologists study these people to fill in the gaps about what is known about early humans. The documentation of their lifestyle by contemporary scientists has shed light on how the earliest people might have lived.

By studying the impressions left by the brains on the inside of skulls, scientists are reaching new conclusions about the brain development and thinking processes of primitive peoples because this new avenue of exploration provides them with knowledge as to when various parts of the brain developed and how they evolved. As current understanding of the workings of the brain grows and the relatively new field of paleoneurology expands, humankind will benefit from this increased knowledge and understanding.

Other Evidence That Provides Clues

The primary evidence studied by paleopathologists is the remains of those who lived previously, but these scientists also look for other ways to uncover information about ancient peoples' lifestyle and health. Art, archaeological evidence, and microscopic clues from waste sites and food preparation areas as well as the insects these people attracted are also very helpful for study. Art always offers insight into how people lived, but two obstacles arise from using it as a method of learning about prehistoric people. The first is limited availability. Art-related evidence from primitive people is in short supply. Prehistoric pottery shards tend to be too small, and the decoration depicted is too incomplete to tell a story. While a few cave paintings exist, there are not many. Since they are relatively recent finds (1940 and 1994), there may be more discoveries as archaeologists more fully explore these new areas and the people who inhabited them. Art also presents another problem. To understand how a particular drawing reflected a people, historians must determine the intent of the artist. Was the artist painting "from life," or was he or she depicting a dream sequence? As archaeologists, historians, and paleopathologists assemble the clues that tell a story of a particular people, they must evaluate whether or not the artist's depiction is true to life or a romanticized illustration.

From the Egyptian pyramids to places like Stonehenge, architectural remains offer information about how the people of a particular day lived and how they died. Paleopathologists owe a debt of gratitude to the pyramid builders since these tombs have preserved evidence of complex burial customs. Because the dead were sent off with everything they would need for life in the hereafter—from their gold to an occasional sacrificed servant or two, the tombs are rich with details of Egyptian life. This information, however, sometimes needs interpretation. Mummies are sometimes found with prosthetic limbs, but scientists are not certain whether these were used during a person's lifetime or whether the prosthetic additions were added to the bodies at burial to provide aid for the person's next life.

Studies conducted by a team of archaeologists from the University of Sheffield have led to the theory that Stonehenge, which dates back 4,600 years to Neolothic times, was one of a pair of locations where people were sent for healing through prayer. Two miles (3.2 km) northeast of Stonehenge, an archaeological team has discovered nine homes in Durrington Walls, built to face the midwinter sunrise (as opposed to the midsummer sunrise faced by Stonehenge). Both locations feature matching roads that lead directly to the river, and it would have been easy for people to travel to the water, board a boat, and move to the "sister" location. A visit to Stonehenge may have been for fertility, or it may have been a healing center. Professor Timothy Darvill, a leading British archaeologist, published *Stonehenge: The Biography of a Landscape* that documents his findings and notes that the skeletal remains around Stonehenge show higher than normal evidence of disease. In addition, some of the stones used at Stonehenge were brought from a distant area (160 miles [257.5 km] away) that was known as a healing center, further supporting this current hypothesis.

Stonehenge is thought to have been a healing center. *(Frédéric Vincent)*

"It was believed that these particular stones had many healing properties because in Preseli [where the stones were from] there are many sacred springs that are considered to have health-giving qualities," said Professor Darvill.

Tiny clues from the past sometimes reveal as much as large archaeological remains. Preserved insect parts or parasite eggs can help scientists determine the diseases that may have been prevalent during a certain period. By identifying that a particular insect existed in prehistoric times, scientists can determine the diseases that might have been common at a specific time. Insects are the carriers of a number of diseases that plague the world today (malaria, West Nile, and Lyme disease, to name just a few), and because scientists have been able to find and date insects from ancient times, they then deduce what illnesses probably existed in that period. When scientists found fossilized tsetse flies from the Tertiary period (more than 2 million years ago), for example, they were led to the likely conclusion that tsetse flies likely spread "sleeping sickness" then as well.

Food preparation areas and waste sites provide additional information. Scientists have found that by scraping microscopic particles from shards of pottery found in cooking pits, they can analyze these samples under a microscope to learn more about the diets of prehistoric people. The contents of cesspools, latrines, and refuse piles are also extremely helpful. Scientists may take for further study fossilized bacteria, pollen grains, seeds, charcoal, hair, bits of bones, shells or feathers to learn what they can about early people.

Testing

As new devices are introduced for testing medical patients of today, scientists soon use the tools to glean new information about ancient specimens. Paleopathology now uses many types of testing, from visual examination to X-rays, computed tomography (CT scans), magnetic resonance imaging (MRI), electron microscopy, and immunologic, chemical, and mass spectrophotometry methods. Early human remains have now undergone CT scans, tissue samples (when available) have been examined under electron microscopes, and chemical analyses have been performed on remnants of tissue and bone.

Another recent development (first used for dating artifacts in 1947) has been carbon 14 dating, a method of obtaining more accurate age estimates on organic materials from thousands of years ago. This method of assessing the age of human remains or artifacts has been instrumental in mapping human history. Until the development of carbon 14 dating and other methods that have followed, archaeologists primarily had to determine the age of artifacts based on an association with what else was found at a particular location and what was known about the people and their time period. (See the sidebar "Radiocarbon Dating" for more information on carbon 14 dating.)

As DNA and genetics have changed the way doctors view current-day illnesses, DNA testing is also being used in paleopathology. In specimens in which DNA evidence can be retrieved, scientists are learning even more about ancient peoples.

Scientists are aware that these methods must be used with an understanding of how time or other influences may have altered a specimen. Salt in dirt can leach out the calcium from a skeleton, leaving scientists to wonder whether the person suffered an illness that reduced bone calcium or whether the calcium had dissipated over time from natural causes, environmental damage, or mistreatment of the specimen.

SPECIFIC ILLNESSES AND INJURIES THAT PLAGUED EARLY PEOPLE

The environment in which primitive people lived was harsh and dangerous, but their lifestyles helped them avoid some types of health problems that plagued people who were to live later on. Hunter-gatherers did not stay in one place long enough to pollute water sources. This practice eliminated heath risks of tainted water as well as diseases that can be spread by insects living near polluted rivers or lakes. Because they had not yet developed the skills necessary to tame and raise animals for community use, domesticated animals were eliminated as possible sources of disease. Because hunter-gatherers lived in low-density groups and did not intermingle with others, contagious illnesses were also rare.

The primary health risks of this time were from injuries or broken bones that could result from being a very mobile population. Most of these injuries healed poorly and probably left the victim living out a life in pain. These early people were also subject to infections from the animals they hunted.

Dental abscesses also may have affected almost everyone. Because people in many cultures used their teeth as tools and would not have realized the importance of dental care, any tooth problem would almost certainly have been painful and would have created hardship.

Actual infections would have varied from civilization to civilization, but the diseases that affected hunter-gatherers primarily came from two types of sources: zoonotic illnesses (diseases carried by wild animals) or parasites and bacteria. Zoonotic illnesses are acquired by humans through exposure to animals or by eating raw animal flesh, which can release microbes that cause relapsing fevers such as typhus (a louse-borne disease), brucellosis, and hemorrhagic fever. Diseases such as anthrax and rabies may have been passed to humans by exposure to predatory animals such as wolves. Specific to primitive people in the New World were such animal-borne diseases as trichinosis and tularemia as well as Rocky Mountain spotted fever and rickettsial disease. In South America, prehistoric people also encountered mucocutaneous

RADIOCARBON DATING

Radioactive carbon 14, an isotope that has become useful in dating ancient artifacts, was first discovered in 1940 by Drs. Martin Kamen and Samuel Ruben, both of whom were chemists at the University of California at Berkeley. The discovery of carbon 14 occurred at Berkeley when Kamen and Ruben were working to find a radioactive isotope of carbon that could be used as a tracer investigating chemical reactions in photosynthesis. (Almost all carbon atoms in nature are carbon 12, containing 12 protons and neutrons in the nucleus, so the identification of this radiocarbon isotope with 6 protons and 8 neutrons was noteworthy.) In 1947, Willard Libby, a chemistry professor also connected with the University of California at Berkeley, was experimenting with different ways in which this isotope could be useful, and he discovered that carbon 14 could be used as a method for dating artifacts and human remains from long ago.

Over time, Libby observed that during photosynthesis, plants absorbed carbon 14, a radioactive isotope (atoms with one of two or more atoms having the same atomic number but different mass numbers). Throughout its lifetime, a plant absorbs a constant amount of carbon 14. When the plant

leishmaniasis (uta), a protozoan disease transmitted by bloodsucking sandflies, as well as Chagas's disease and others.

Hunter-gatherers were also commonly plagued by another type of illness, those that came from parasites and water- and soil-borne bacteria. The camplike living conditions of early humans and the need for frequent visits to water sources would have made contact with these minute or microscopic organisms unavoidable. Parasitic worms and lice and bacteria such as *Salmonella* and *Treponema* (agent of yaws and syphilis) were common, and soil-borne

dies, it stops absorbing this radiocarbon element, and the rate of decay of the carbon 14 is predictable and measurable in objects as old as 45,000 to 50,000 years. This measurable rate of decay is known as an object's "half-life." Libby discovered that by determining the concentration of carbon 14 left in the remains of a plant, he could calculate how much time had passed since the plant died. Since animals and humans eat plants, they also undergo an even rate of decay when they die. Because most artifacts are made from an organic substance, the date of the object can be judged, based on the decay of the organic content. This finding has enabled more accurate estimates of the age of many artifacts and skeletons that formerly were dated through less reliable methods.

Because objects more than 45,000 to 50,000 years old do not have enough carbon 14 to measure, scientists have found that they can use similar dating techniques by employing elements with a longer half-life (time to decay) than carbon 14. (Among the elements most widely used currently are potassium 40 with a half-life of 1.25 billion years, uranium 238 with a half-life of 4.5 billion years, and rubidium 87 with a half-life of 49 billion years.)

bacteria could have entered skin wounds and produced gangrene or botulism.

Early humans may have also suffered from malaria or yellow fever, but because the population density was low, these diseases would not have tended to spread very actively. In some cases, people seemed to have developed a tolerance for certain types of disease just as they do today. Sometimes, immunity was passed by mother to baby through the placenta. In other cases, the lack of susceptibility may have been genetic. Individuals with the sickle-cell trait, for example, are not susceptible to the most dangerous form of malaria, and that seems to have held true then as it does now.

Early Treatments

Primitive people had no knowledge of germs and the role they played in causing illness. Sicknesses were attributed to gods or to curses. Prehistoric people sought the best remedy they could think of, based on what they knew. Treatments would have often relied on a combination of religious beliefs and practical remedies.

Medical treatment generally was administered by a shaman or a medicine man, the person within a community who was believed to be empowered by the gods. This esteemed member of the group would provide a drink, offer an herbal cure (in the form of something to eat or an herbal salve on a wound), or say a prayer, and then perform a dance or say an incantation. Cave paintings in France that date from 17,000 years ago show art renderings of masked men wearing animal heads performing ritual dances. (These may be the oldest surviving images of medicine men.) Later, as communities grew, there might have been a division of labor among healing duties. Medicine men might have been divided by specialty, ranging from diviners and shamans to birth attendants and witch-smellers (those who could locate and banish a disease-causing witch).

After an ailing person had consulted a medicine man, he or she likely would have been cared for collectively by members of the community, anthropologists maintain. Healing would have assumed a high level of importance. Most tribal groups could not have afforded to lose a contributing member, and if a person were sick or lame, he or she would have slowed the pace of the tribe's following its food sources. Infant mortality was likely to have been high because there would have been few times when a group could have halted for an ill baby. The birth rate would have also been low since nursing would have been the best method for feeding an infant, and nursing usually inhibits ovulation, thereby preventing pregnancy.

Many populations became adept at caring for open wounds or broken bones. Treatment of an open wound might have included cleaning and packing it with extracts from plants or part of a plant, some of which might have been helpful in cleansing or healing the injury. Cuts were treated with animal fat. Sometimes, animal excrement was rubbed on the wound, and animal skin was used as a bandage.

A broken leg or arm was covered in river clay or mud. This formed a cast of sorts that was hardened by the Sun. Whether the bones in these casts were actually set in place is debatable. Skeletal remains indicate that people often recovered following a bone break, but recent thinking is that good fortune as much as stabilization of the limb may have been what enabled recovery to take place. A study of gibbons, conducted by anthropologist Adolph Schultz, showed that the healing rate of bone breaks of the gibbons and of primitive humans was about the same.

Primitive people's anatomical knowledge was very slight. Although cave paintings show that they knew the location of the human heart, it seems that little else was understood. Any knowledge they had, in any case, did not necessarily result in better health care. Even when they would cut into a body part, such as for an amputation, the process did not necessarily advance their medical knowledge as they did not yet understand the inner workings of the body.

Medicines

Pharmacological knowledge came from experimentation. A good number of the herbal medicines were likely effective, as they are

The eyebright plant was used to heal diseases of the eye because it resembles an eye.

the basis of some of the pharmaceuticals in use today. Early people, unfortunately, almost certainly located poisonous plants some of the time as they sought to sample and identify ones that might have had healing properties.

Primitive people sometimes judged the use and purpose of a plant by examining what the plant resembled. The plant, eyebright, was used for diseases of the eye because a black speck in the flower looked like the pupil of the eye, according to William Osler in his *The Evolution of Modern Medicine.* (Osler was a 19th-century physician who wrote extensively about medicine.) A plant with a bright yellow flower was used in an effort to rid a person of jaundice, which can turn the white parts of the eye yellowish in color.

Early Attempts at Disease Prevention

Anthropologists have observed a few of the primitive societies that have existed in more recent times, and these experts note that prehistoric people often carried amulets as a way to guard against disease. Ritualistic mutilations such as circumcision and *scarification* (intentional creation of scars) were performed for similar reasons. Later on, scarring was used as the entry point for snake venom or smallpox varicella in a very early method of vaccination. Drinking the blood of warriors was also thought to give strength for healing.

Hiding excrement was also believed to help maintain good health. If a person's excrement, clipped nails, or locks of hair were carefully disposed of, then a sorcerer could find no bodily "waste" on which to cast a spell.

TREPHINATION: A COMMON FORM OF SKULL SURGERY

Primitive people had few surgical tools and likely knew little of either anesthesia for pain relief or antiseptics for cleanliness; yet one of the medical procedures that they commonly performed was skull *surgery,* known as *trephination* or *trepanation.* (The word derives from the Greek *trypanon,* meaning "auger" or "borer.") The process of trephination involves removing a small, circular piece of the skull from a patient to expose the *dura mater* (fibrous membrane forming the outer envelope of the brain). While the risk of infection, brain swelling, or bleeding would have been high, perhaps the fact that only a small circle of bone was removed may have been key to the number of people who actually survived the procedure.

Trephination was widespread and common within many communities, and evidence of this type of surgery appears on skulls that date as far back as 10,000 years ago. The procedure was performed on men, women, and children, and some skulls have multiple holes. Ancient skulls with signs of trephination have been discovered in Europe, North Africa, Russia, Bolivia, the Canary Islands, and Peru. At one site in France where the artifacts date to 6500 B.C.E., 120 prehistoric skulls have been studied, and 40 of these show signs of trephination.

Trepanning was performed in slightly different ways in various parts of the world. Primitive people everywhere would have used a sharpened rock, likely flint, as the cutting tool. Depending on the culture, several styles of cutting would have been used. Some people used a scraping method to create the circular opening; others made tiny holes to create a perforation, which might have made the disk of bone easier to remove. Only later did metallurgy bring about the option of using metal tools.

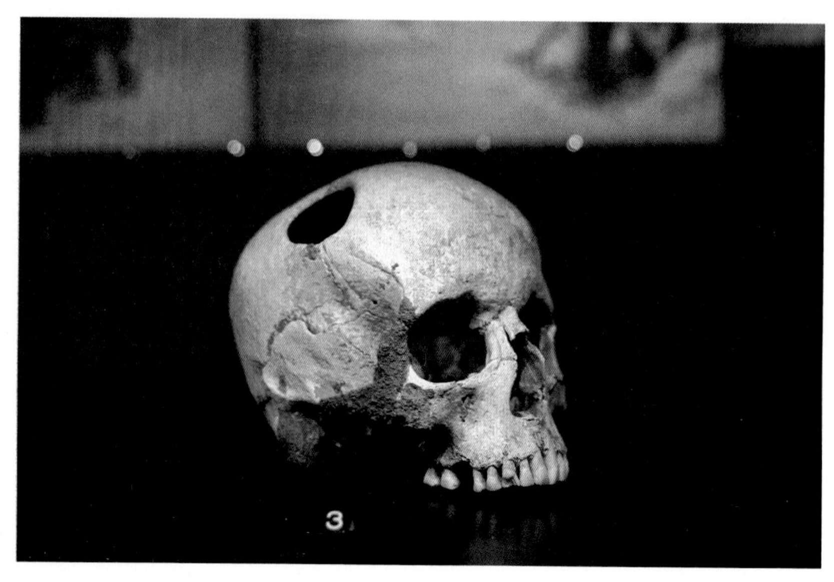

A female skull from the Neolithic era (3500 B.C.E.); the patient survived, as there is evidence of healing. *(Natural History Museum, Lausanne)*

Modern people did not become aware of the primitive practice of trephination until the 19th century. Ephraim George Squier (1821–88), a U.S. diplomat-journalist, acquired a primitive skull from Peru. In 1865, Squier, also an amateur anthropologist, showed others the skull. It had a well-defined hole, but because there were signs of new bone growth (healing) around the hole, Squier's contemporaries realized that the procedure was done with enough care that the person actually survived. This led to a new investigation of other skeletal remains, and scientists soon realized that it was a commonly performed procedure.

The reasoning behind this type of skull surgery likely varied with the culture. Headaches, infections, convulsions, or signs of insanity may have led to trephination as a way to relieve suffering. Because of the frequency with which it appears to have been performed, paleopathologists speculate that some groups may have used it as part of a magical or religious ritual to release evil spirits. In many cultures, the disklike fragments of bone that were removed from the skull were likely worn as amulets (good-luck charms), which further bolsters this theory.

In Peru, the process often followed the occurrence of a skull fracture. If head trauma was the reason for trephination, then a trained medical person performed the surgery. If the surgery was being done for religious or supernatural reasons, however, then a shaman performed the procedure.

What was done to relieve pain during surgery is unknown. Liquor may have been offered the patient, or in South America, it was likely understood that coca (used to create cocaine) could function as an anesthetic. Occasionally, there are signs that the procedure was begun and then halted. Perhaps it was simply too painful.

Many years later, Hippocrates wrote of the process as it was performed in Greece, and it continued to be practiced through the Middle Ages and the Renaissance. By that time, it was used as a method to cure seizures and skull fractures. The survival rate continued to be quite good. Today, this procedure is still used occasionally for epidural and subdural hematomas and for surgical access for certain other neurosurgical procedures.

Other Forms of Early Surgery

Prehistoric people also performed other types of surgery. Most procedures were noninvasive and were designed to relieve swelling or pressure, such as the lancing of boils. Early people also undertook the amputation of injured arms or legs, probably in an effort to save lives. (Amputation of fingers is thought to have been a religious rite.) While survival was by no means guaranteed in an amputation, the examination of a skeleton at the site of an amputated limb often reveals healing, so people did survive.

These early "surgeons" usually restricted themselves to operations on limbs. They must have recognized that surgery in the abdominal area put the patient at high risk. Until about 1850, any surgical procedures needed to be brief because there were few ways to numb the pain. Anesthesia did not exist, so people were limited to whatever calming plant or drink was available to them.

Because of the emergency nature of some cases of childbirth, experimentation to create a method to perform cesarean sections took place in ancient times. The goal generally was to remove a living infant from a mother who had died. Successfully operating to save the life of the mother was impossible at that time.

THE DISCOVERY OF THE ICEMAN

In 1991, when German tourists Helmut and Erika Simon came upon an old frozen corpse in the Ötzal region of the Alps between Austria and Italy, they knew to contact authorities immediately, but they had no idea of the importance of their find. The authorities did not respond with any particular urgency, so the body was left exposed for a few days where it was chipped at by skiers who passed by. By the end of the week, forensic doctors, instead of paleopathologists, came to remove what they thought was a 500-year-old specimen. Only later did scientists discover that the body, which had been removed from the ice rather roughly, was actually a naturally preserved mummy from approximately 3300 B.C.E.—more than 5,000 years old.

Despite the casual handling of the discovery and some heated disputes over "ownership," Ötzi, as he is called, has provided a new level of knowledge about ancient peoples' past, their health, and their way of life. Although the body has been carefully studied by a variety of scientists and medical professionals for more than 16 years, advances in science have meant that the thinking about his health, his diet, where he was from, and his cause of death have altered over time. What scientists have learned and how they have learned it is quite fascinating.

Studies of the composition of his tooth enamel reveal that Ötzi grew up near the present-day Alpine village of Feldthurns. By examining his hair, they have determined variations in his diet during a period of time. After snipping out an extraordinarily small sample of his intestine (a minute bit of his colon), scientists were able to identify the contents of his previous two meals, the last of which had been eaten eight hours previously. One meal involved chamois meat, the other red deer meat, and both meals were eaten with some grain as well as roots and fruits. The grain was highly processed wheat bran, and it may have been eaten in the form of bread.

Based on the pollen in his system, scientists determined that he died in the springtime and was in a midaltitude conifer forest when he ate his last meal. Because he had recently eaten sloes (small plumlike fruits of the blackthorn tree), experts know that early people of that time must have understood how to store and preserve food from season to season. The fruit would have been harvested in the fall, and the wheat would have been grown the previous summer.

Even Ötzi's cause of death took a long time to identify. While it was originally believed that he might have frozen to death in an unexpected spring snowstorm, a recent (autumn 2006) discovery led to a new conclusion: Ötzi died from a fatal battle wound. Because of the careless way in which he was extracted from the ice when he was originally found, Ötzi's clothing had been damaged. In a follow-up study of the body conducted quite recently, one of the workers noted a previously unobserved hole in the shoulder of his garment. This led to the discovery of an unnoticed injury that showed that an arrowhead must have pierced his left subclavian *artery.* In 2007, researchers from Switzerland and Italy used newly developed medical scanners to examine the hunter's corpse, and they confirmed the earlier finding that the arrow had torn a hole in an artery beneath his left collarbone, leading to a massive loss of blood. That, in turn, caused Ötzi to go into shock and suffer a heart attack, according to an analysis published in the June 2007 issue of the *Journal of Archaeological Science,* which was cited in the *Washington Post* on June 11, 2007. (But just to show how quickly new opinions evolve, the *Los Angeles Times* of September 1, 2007, reports that unnamed radiologists, pathologists, and other researchers, using new forensic information and CT scans, say that they believe that blood loss from the arrow wound only made

Ötzi lose consciousness. They say that he died either from hitting his head on a rock when he passed out or because his attacker hit him in the head.) Further examination of the remains showed blood of others on his garments. More blood was found on the left side of his clothing, indicating that he may have been traveling with one or more people and that he may have been carrying someone who was injured just before he died.

Studies of Ötzi's state of health indicated that he had whipworm, an intestinal parasite that has plagued people throughout the ages. His body bore 57 carbon tattoos, simple dots and lines on his lower spine, behind his left knee, and on his right ankle. X-rays revealed that he may have had arthritis in these joints, so whether these tattoos were from some form of acupuncture, possibly used for pain relief, or whether they signified some other rite of passage is not presently known.

Ötzi carried with him two types of polypore mushrooms strung together on leather strings. One of these, a birch fungus, is known to have antibacterial properties, so he likely carried it for some type of medicinal purpose. The other fungus carried was a tinder fungus, which would have been helpful in starting a fire, along with other items that were part of what would have been a fire-starting kit (flint and pyrite for creating sparks). He also had with him more than a dozen different plants. He and his contemporaries must have believed that they had medicinal benefits.

MESOPOTAMIA AND THE BEGINNING OF HISTORY

At the end of the last Ice Age (about 10,000–12,000 years ago), the world population consisted mainly of hunter-gatherers, but here and there, civilizations were beginning to develop. Particularly in areas along riverbanks where water was accessible and the land was fertile, some early groups began to discover that they could grow grain that could feed a small community. As farming developed, the lifestyle of these people underwent profound changes. A nomadic life was no longer necessary, so people could build towns and villages and live near one another. Because not everyone had to be devoted to the gathering or growing of food, some people were free to create tools, build places to live, weave baskets, and otherwise contribute to community well-being.

By the Bronze Age (ca. 4000 B.C.E.), people had become quite skilled at metalworking, which simplified much of the work that had to be done in the communities. In addition, a calendar system was invented, enabling more accurate prediction of the times when riverbeds overflowed their banks. Knowledge of the cycles of river levels allowed water to be properly channeled, thereby helping with crop irrigation. During this period, people enjoyed a more stable lifestyle, and as a result, they had time to pursue other skills, among them a method for writing. They began to record information about their lives and about their healing methods. At this point, history began.

Mesopotamia, the area between the Tigris and Euphrates Rivers also known as the Fertile Crescent (part of present-day Iraq),

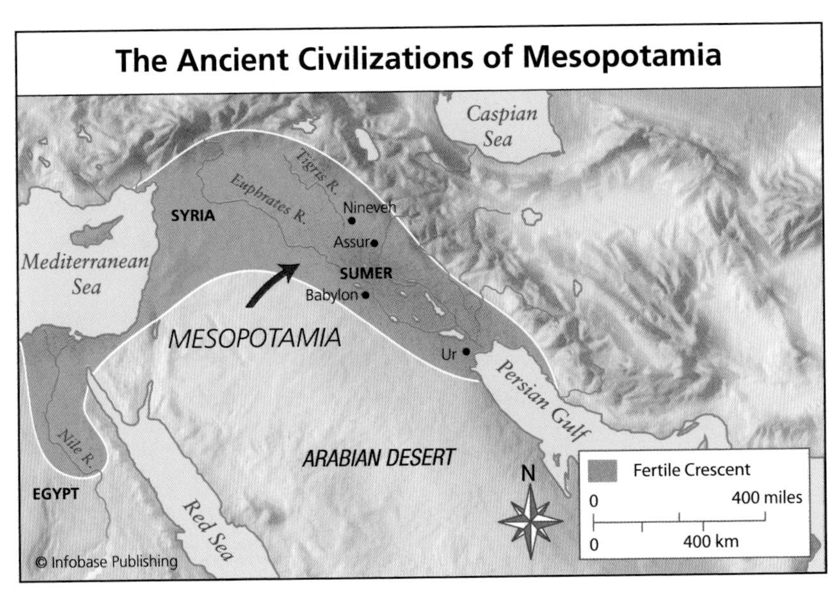

The area between the Tigris and Euphrates Rivers, known as the Fertile Crescent, is the site of an area that was particularly beneficial for farming. As a result, cities such as those indicated, rose in this region, which is now in present-day Iraq.

has long been known as a prime location of early farming, as well as the site of several of the earliest civilizations in the world. The Sumerians gave birth to the first true civilization, the city of Ur on the Euphrates, which lasted from about 3300 B.C.E. to about 2000 B.C.E. Babylon was located further up the Euphrates; Assyria was settled on the Assur, while Nineveh grew up on the Tigris River. All these settlements made lasting contributions to development in Mesopotamia.

Though for the most part, agriculture improved the people's lifestyles, it also created new health problems. Early farmers, for example, suffered from illnesses because their bodies were missing certain nutrients. Because they grew only a few types of grain, their nutritional needs were often unmet because of the lack of variety in their diet. As it happened, the forced variety of diet that was necessitated by hunter-gatherers needing to eat what they could find actually had provided better nutrition for the farmers' forefathers.

DISEASES THAT THRIVED IN CIVILIZATION

Major infectious diseases that are able to pass from person to person all seem to have originated with animal hosts. Smallpox is thought to have appeared as early as 3000 B.C.E., and pox-type illnesses were brought to the human population from cattle as more and more cows became domesticated. Cattle also brought tuberculosis; pigs and ducks both carried influenzas; the domestication of horses exposed people to rhinoviruses and the common cold. Measles jumped from dogs (where it appeared in a canine distemper form) to cattle and then to human beings.

Disease-carrying insects began to hover near the waterways that were being polluted by animal and human waste. Water in areas that became heavily settled became polluted with animal feces that spread polio, cholera, typhoid, viral hepatitis, whooping cough, and diphtheria. (Thousands of years would pass before the dangers of unclean waterways were generally recognized.)

Irrigation and farming also caused an increase in parasites and parasite-carrying disease, among which were roundworm (*Ascaris*) and hookworm (*Ancylostoma duodenale*) from pigs. In areas with irrigation, parasites entered the bloodstream of barefoot field workers. The blood fluke, *schistosoma,* was among the parasites that spread easily.

For the first time, people lived in close enough proximity that airborne illnesses could spread from person to person with no intermediary carrier. Over time, some people built up immunities. The survivors became part of a population's recovery until the disease sometimes returned and further decimated a city.

Medical Treatment in Mesopotamia

The earliest known medical records found are from Mesopotamia. They date to the seventh century B.C.E. and provide historians with a glimpse at medical care in early civilization. Among the more than 30,000 surviving clay tablets, written in cuneiform, about a thousand of these tablets concern medicine. Translation and publication of the medical tablets started only in the 1920s. (The first effort at translating cuneiform was made in 1835 by British Colonel Henry Rawlinson.) Most of the medical tablets are prescriptions; a few describe specific procedures such as the draining of pus from a lung (pleura). One of the main texts that has been found is a treatise, organized head-to-toe by disorder, outlining several centuries of Mesopotamian medical knowledge, including treatments going back further than the seventh century B.C.E. The 40 tablets on which this information appears has been collected and studied by the French scholar R. Labat, and while the cuneiform translations are still being puzzled over, this set of tablets is referred to as *The Treatise of Medical Diagnosis and Prognoses.* Among the information gleaned from the translation of this document is a description of tuberculosis, a cure that involved bleeding (a healing method commonly used for thousands of years), as well as specification regarding what fees should be charged for various medical services.

Prognostication (future-telling) was an important part of medical tradition at this time, and the Mesopotamians relied on various methods, including astrology, astronomy, and hepatoscopy (the study of the shape of the liver of a sacrificed animal). The liver was thought to be the seat of life, so a healer might examine the liver of an animal, or the healer might observe the flickering of a flame to divine whether or not the patient was treatable.

Before undertaking a case, a Mesopotamian healer needed to evaluate a patient to decide whether or not he or she would likely live, since a failure (death) would reflect poorly on the physician. Common practice was to take a look at the patient, evaluate the odds, and simply not treat those who were unlikely to recover. While an experienced physician would be taking into account the overall pallor of a patient and his or her basic demeanor, the physician might also call on other spirits to tell him or her what to do.

A custom of the day in Mesopotamian cities, according to the Greek historian Herodotus, was for the sick to be placed outside of his or her house so that friends and neighbors could give advice about possible cures. Although this practice may have indeed occurred, the Mesopotamians clearly had other methods that they used in healing their fellow human beings. By this time, the healing profession had developed two types of specialists described in Babylonian texts: the *ashipu* and the *asu*. The *ashipu*'s role was to diagnose the cause of the ailment. This often involved deciding which god or demon caused the illness or whether the illness was brought on by a sin of the patient. The *ashipu* could also cure through the use of charms.

The *asu* specialized in wound care, which involved washing, bandaging, and making herbal plasters. Some of the recipes for the plasters show an understanding of effective elements that would have had healing benefits. (One plaster called for heating of plant resin or animal fat with alkali. Alkali creates a soapy substance that would have helped clean wounds.) The two specialists did not seem to compete with each other, although only the wealthy could have visited them both.

Cleanliness was also important to Mesopotamians, and archaeological remains show evidence that they had sewage systems and draining methods. They also understood the notion of contagion with regard to some illnesses, and lepers were among those who were isolated to protect other people.

Further information about medicine in Mesopotamia comes from the Law Code of Hammurabi. Hammurabi was a ruler of Babylon from 1792 to 1750 B.C.E. who will long be remembered for establishing a code of laws that dictated proper governance and laws for the citizenry as well as rules for practicing medicine. These laws were not written on a tablet but on a large block of polished diorite (ca. 1750 B.C.E.), which is currently in the Louvre Museum in Paris. Among the legal dictums that pertained to medicine were laws that held doctors responsible for surgical errors and failures. The laws describe punishment only for medical errors in "use of a knife," so perhaps nonsurgical mistakes did not carry such harsh punishment. The punishments themselves had to do with the status of the patient. If a person of high status died from surgery, the surgeon might have his hand cut off; if a slaved died, the surgeon had to pay to replace the slave. Fees for services were also specified, and doctors were better rewarded for taking care of a person of status than for caring for a slave.

CONCLUSION

Good health was as important in ancient times as it is now. Although relying heavily on religion and magic, primitive people took early steps to learn about plants that had healing properties and to discover practical treatments that helped cure ailing members of the community. Primitive people lived on six continents, and despite a lack of communication among various groups, many of the healing methods they employed were remarkably similar to one another.

Although people had begun to build up some resistance to the illnesses in their own geographic area, diseases began to spread more broadly as people began to travel to other communities to trade information and goods. Marauders, merchants, missionaries, and armies also traveled far from their own native lands, and a disease to which one community might have built up resistance became another community's *plague*.

It would be thousands of years before real answers to preventing or curing disease were to be known, but in all civilizations, people did their best to find ways to stay well. In the following chapter, civilization takes a giant leap forward with all that was accomplished by the Egyptians. Their amazing pyramids, filled with all that an Egyptian ruler might need for the afterlife, and their mummification of bodies to preserve them for life in the next world has permitted a new level of understanding about the health beliefs and medical treatments that were part of the Egyptian culture for almost 3,000 years.

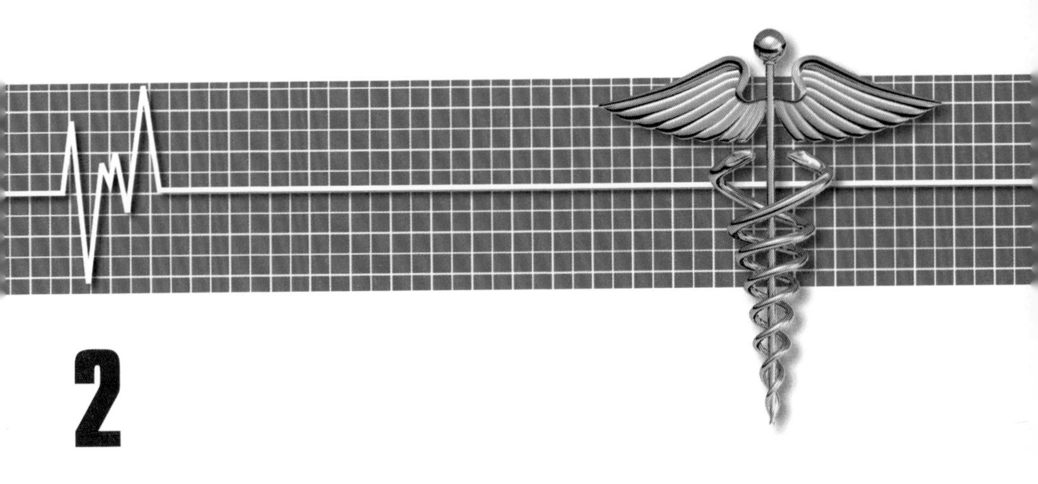

2

What the Mummies Reveal

As early as 11000 B.C.E., settlements began to sprout along the Nile River in the area that was to become Egypt. The people found a plentiful supply of fish and eventually (ca. 5000 B.C.E.) learned that the regular flooding of the river created fertile soil for farming. The Nile also offered the benefit of being navigable in both directions so that by 3200 B.C.E., Egyptians were taking advantage of many trade opportunities up and down the river, though the inhospitable desert meant that they did not venture too far inland.

While the Mesopotamian area of this period featured many different city-states, ancient Egypt initially developed as an Upper and a Lower Kingdom that were eventually united as one (ca. 3000 B.C.E.). From 2575 to 2150 B.C.E. (a period known as Old Kingdom), Egyptian civilization expanded and enjoyed a remarkable 3,000-year period of calm and governmental permanence, partly because of the isolation and resulting protection offered by the broad expanse of desert that surrounded the settlements.

The achievements for which ancient Egypt is recognized blossomed during this period of stability, with advances in building, technology, medicine, hieroglyphic writing, and artistry. While the creation of the solar calendar (based on observations of the regularity of the flooding of the Nile) is an oft-mentioned, lasting contribution of ancient Egyptians, there were also developments that had direct bearing on the people's health and, therefore, on the medical practices of the day. From the building of the pyramids to the hot, windy weather to the types of bread they ate that caused cavities, the Egyptian people suffered from illnesses and injuries that were unique to their life experiences.

As with other ancient cultures, some of the knowledge of ancient Egypt is based on information noted by writers from a different age. Among those who described Egyptian medical practices in their work were Homer, in *The Odyssey,* (800 B.C.E.), Herodotus (the Greek historian who visited Egypt in about 440 B.C.E.), and Pliny the Elder (23–79 C.E.), whose writings included *Naturalis Historia,* which includes information about ancient medicine from Egypt. But because the Egyptians created a useful form of picture-writing known as hieroglyphics, these documents—finally translated in the late 19th and early 20th centuries—have provided people of the future with a lasting testament about Egyptian life and the medical beliefs of the day.

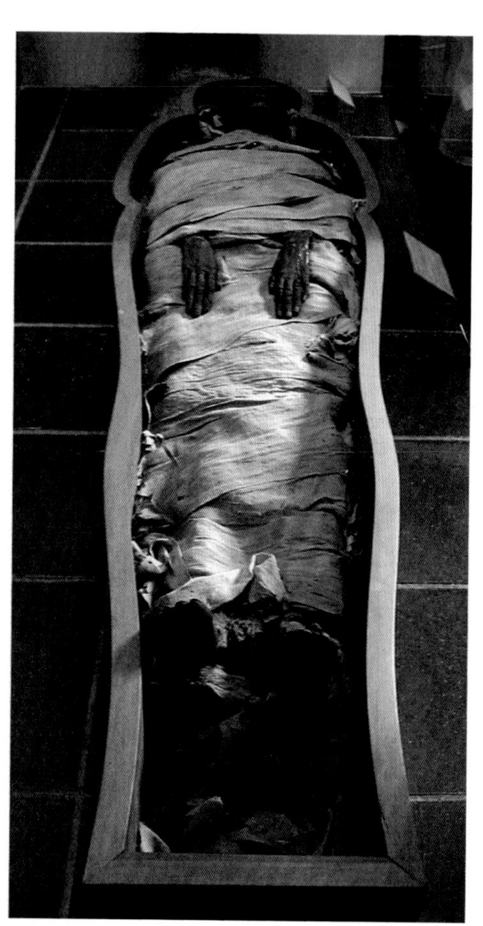

Egyptian mummy that is kept in the Vatican

Much of the initial medical writings were collected in the library established in the city

of Alexandria. This major Egyptian port city was founded by Alexander III of Macedon (also known as Alexander the Great, 356–323 B.C.E.) in 323 B.C.E., and the library was established by Alexander's half brother, Ptolemy. Both the library and the School of Medicine that was located nearby were oases for scholars until a fire occurred in 391 C.E. Many documents were lost at this time, and those that survived were scattered and refound at much later dates.

Ancient Egyptian medicine dates from about 3300 B.C.E. until the Persian invasion of 252 B.C.E. While the Egyptians came to rely more and more on magic, many of their early medical beliefs were firmly rooted in rational thought. They also successfully gathered an amazing amount of knowledge about healing plants and contributed to the field of pharmacology. This chapter will explore the process of mummification and Egyptian dentistry, and it will introduce the Egyptologist who created the field of paleopathology.

EARLY EGYPTIAN MEDICAL BELIEFS

Each group of people has their own unique explanation for the cause of illnesses. The early Egyptians (ca. 2500 B.C.E.) believed that everyone was born healthy, but evil spirits sometimes entered the body and needed to be banished through prayers, magic spells, and magic potions. Because people often begin to feel better on their own, these magical methods often seemed effective. Egyptian healers also learned how to create medicines, many of which earn nods of respect from today's medical community. (These medications will be described later in the chapter.)

The Egyptians had some degree of understanding of the circulatory system. They knew that the heart was the key to life, and they felt that all bodily fluids, from blood and urine to sputum and solid waste, moved through the body's vessels in a manner similar to the flow of a river like the Nile, the natural entity that influenced so many other aspects of their lives.

The Egyptians were plagued by illnesses that had to do with their lifestyles. The unremitting sun and wind caused eye ailments, and the workers involved in the building of the pyramids were likely to suffer bone breaks. Tooth problems were also common.

The Egyptians recognized the need for sanitation and cleanliness (bathing and shaving under the arms was stressed). While their city system was not as advanced as Rome's would one day be, they did have water closets in the home, and waste water was carried out of houses in copper pipes. They also slept under mosquito nets, so they either found the insects very annoying or they had begun to suspect that these insects carried disease.

Surgery for amputations involved the use of a knife made of flint or obsidian (a black stone sharper than today's modern stainless steel blades) before iron and bronze. Once metal blades were used, incisions could be cauterized. Ancient texts explain that if the blade was heated before the incision was made, then the edges of the wound were sealed as the surgeon made the first cut.

Before Greek physicians took on supreme status as the best doctors available, Egyptian physicians were summoned to other countries to treat the sick who could afford to summon an expert healer. There were three categories of healers who might have been consulted. According to the Greek historian Herodotus, these included physicians, exorcists, and priests of Sekhmet (thought to be surgeons or specialists in feeling the *pulse* and treating diseases of vessels). Egyptian medical practitioners were trained in temple schools and were considered part of the priest caste. The Egyptians also relied on many different specialists, including a much written-about proctologist (*neru phuyt*), which sometimes translated as "shepherd of the anus." While the heart was viewed as the center of life, the anus was viewed as the main seat of pathology, so this specialty would have been a very important one. Other specialists included those who focused on diseases of the eye, the head, the teeth (early dentists), and the belly. The need for the early equivalent of veterinarians had become apparent, and there were even pest control officers who were responsible for ridding houses of fleas, lice, rats, and other types of rodents and bugs.

A few women were also a part of the medical system. (Women in ancient Egypt could own property and had a good number of legal rights.) They served as nurses, wound dressers, and sometimes as physicians. The female physicians tended to health needs of women as well as overseeing the use of health and beauty treatments involving skin creams, ointments, and hair dyes. A tomb that dated to 2400 B.C.E. notes that it held the body of Lady Pesheshet (Fourth Dynasty), who was "lady overseer of the lady physicians." She is thought to have been the first female doctor.

Imhotep, the Egyptian who was to be elevated by the Greeks to the status of the god of medicine, served as a key adviser to Pharaoh Djoser (also known as Netjerikhet), who ruled during the Third Dynasty (2686–13 B.C.E.). Because of Imhotep's exalted position in the direct employ of the pharaoh, he worked in many arenas besides medicine. One of the most lasting contributions he made was in his work as designer of one of the earliest pyramids, known as the Step Pyramid. Before the existence of the pyramids, pharaohs were buried in simple mastabas (mud-brick tombs with flat roofs and sloping sides) in mortuary complexes. For Pharaoh Djoser, Imhotep kept adding mastabas in a pyramid-like arrangement to create a grander and grander burial place. This pyramid eventually became a six-step pyramid more than 200 feet (62 m) high.

The healing methods recommended by Imhotep (his name means "one who walks in peace") were cures achieved by "incubation." At the time, sleeping—and dreaming—within a healing temple were believed to be curative, so patients would come and sleep at his temple, meet with one of the attending priests in the morning, and then believe themselves cured afterward. During Roman times, Imhotep's tomb in Sakkara was considered a shrine.

MEDICAL PAPYRI

The oldest existing medical documents are Egyptian papyri that were discovered near Luxor, Egypt, and purchased by Edwin Smith in 1862. One, sometimes described as a 110-"page" scroll, is a surgical papyrus that documents 48 different battlefield wounds and how to treat them; it is referred to by the name of the purchaser. Another, the Ebers Papyrus, which explains more about disease management, was sold in 1872 to Georg Ebers, a German Egyptologist and writer, and it bears his name. (Other papyri are named after where they were found.) Though there are other medical papyri, the Smith and Ebers Papyri are generally considered to be the most significant because of their age and the volume of information that they present. These papyri, particularly the Smith Papyrus, reveal a relatively high level of medical knowledge and show that *empirical* (experience-based) treatments were used in addition to magic. (For more information on the translation of these papyri, see the sidebar, "The Importance of Papyri and Related Developments.")

Edwin Smith Papyrus

Sometimes referred to as the "Book of Wounds," the Smith Papyrus dates to approximately 1600 B.C.E. and contains descriptions of 48 different medical cases involving the treatment of various types of wounds. Because there is reference to medical information that dates to the time of Imhotep (2640 B.C.E.), it is thought to be copied partially from even older documents.

The wounds and the recommended cures are organized by body part. Because only a portion of the document remains, the existing information starts with head injuries but includes nothing about any body part lower than the thorax. The treatments described range from methods that were probably quite effective to those that would have been less so. Among the recommendations was one for placing raw meat on wounds. This is a method that could have been helpful in stopping bleeding. Honey was also thought to be curative, and today it is recognized as a potent hygroscopic material (absorbs water) that would have stimulated the creation of white blood cells. Sour or moldy bread was also used for wound treatment; this is particularly fascinating since penicillin is extracted from moldy growths. (It took until 1928 before anyone discovered a way to create penicillin from mold.)

Other methods may not have worked so well. A headache cure noted here involved rubbing the head of a patient for four days

The Edwin Smith Papyrus features advice relating to trauma surgery. *(Jeff Dahl)*

with a poultice made of "skull of catfish." Since trephination was so popular in so many early cultures, the absence of a description of this type of skull surgery is significant. For some reason, Egyptians must not have used this type of cure.

Within the physician class, the following were several levels of professionals (and are explained in the Edwin Smith Papyrus):

- "Swnw" (pronounced soonoo), doctor of the people. This position was the lowest of the medical people. These people were usually self-taught, learning mostly through trial and error. There were both junior doctors (swnw) and senior doctors (wr-swnw). They also sometimes performed dentistry.
- "Wabw" (pronounced waboo), meaning the pure. This physician was of higher status and dealt with the more

privileged in society. They administered medicine, but religious rites were an important part of their duties.

- "Saw" (pronounced saoo) means guardian; their practice was limited to royalty and other elite members of society. They gained education and training within the temple palace schools and also underwent formal apprenticeships. They used the techniques of the wabw but add in magic and sorcery. They were the specialists who treated the pharaohs and would also have served as health ministers on an as needed basis.

Even with a range of specialists, many Egyptians would have had to rely on family and friends for health cures. Locating an

THE IMPORTANCE OF PAPYRI AND RELATED DEVELOPMENTS

While the Mesopotamians were etching cuneiform symbols into bulky clay tablets, the Egyptians were delighting in their discovery of papyri, a very light, sturdy, and economical material on which one could write. Using the native-grown Nile valley papyrus plant, Egyptians discovered that if they placed thin strips of the pith (spongy tissue in the stems) together and soaked, pressed, and dried them, the resulting product would create a sturdy surface on which to make notations. The sturdy quality of the papyri and the dry climate in Egypt has helped preserve these documents.

As functional as this invention may have been for the ancient Egyptians, the papyri documents with their strange picture-writing proved baffling to explorers until the 19th century, when a discovery was made that opened the door to the possibility of translation. In 1799, the Rosetta Stone was unearthed by French soldiers working in Egypt. The stone

affordable and available physician would have been difficult for the lower classes.

Ebers Papyrus

The Ebers Papyrus dates to about 1550 B.C.E. and emphasizes many remedies for getting rid of demons. It also refers to the long tradition of observation and empirical practice (experience-based theories based on observation but not necessarily based on scientific reasoning). This document begins with three incantations to be used when giving remedies or removing bandages, but after this general material, it is broken into multiple books, including specific volumes on internal diseases and eye and skin ailments. Depression and "sickness of the heart" are

featured writing in two languages (Egyptian and Greek), and it used three scripts (hieroglyphic, demotic, and Greek). Though unraveling the meaning of the hieroglyphics on this one item took 23 years, it eventually unlocked the Egyptian written language and provided the key to translate other documents, a good number of which were totally devoted to medicine.

Two of the major medical documents were not translated until much later: The Ebers Papyrus, currently in the University of Leipzig, was translated in 1890, and the Smith Papyrus, eventually donated to the New-York Historical Society, was translated within a period of 10 years (1920–30) by James Henry Brested. The completion of the Smith translation was particularly significant as the medical advice it contained was less magical than that of the Ebers Papyrus, so it provided a new level of respect for Egyptian medicine.

addressed separately as are obstetrics, contraception, and dentistry. The eighth book is about *anatomy* and *physiology* with a good description of the circulatory system. The ninth volume addresses surgery.

In addition to describing various ailments, the Ebers document also reflects 876 prescriptions made from more than 500 substances, ranging from minerals such as lead and copper salts to vegetable matter such as gentian, senna, castor-oil seed, and pomegranate, as well as animal substances.

The Ebers Papyrus outlines the following four-step method for conducting a clinical exam:

1. Preliminary diagnosis (prognosis). Physicians took an early opportunity to decide whether or not to treat a patient. If the person was already very sick or if the disease itself seemed incurable, then the physician avoided taking the case. It was important to maintain a good reputation, and to do so, physicians found it necessary to refuse to treat those who were very likely to die anyway. (This was common practice in many cultures until the 18th century.)
2. Examination of the patient. The Egyptian physician used all five senses in making an evaluation of someone who was sick. He would check the patient's pulse rate and palpate (feel) male patients as well as taste and sniff the patient's bodily fluids—from sputum (spit) to urine and blood. (The vast majority of physicians were male, and a woman who was ill would not be examined by a male. Her symptoms would more likely be learned via a verbal report through an intermediary, who was unlikely to have been a physician.)
3. Diagnosis. The physician would announce an opinion on the patient's condition.
4. Plan for treatment. Recommendations of the appropriate therapeutic measures, which ranged from medicines and manipulation to magic formulas and prayers.

Because physicians tried to preselect patients who would recover, their success rate was better than it would have been if they were willing to take on any patient who was sick.

Other Significant Medical Papyri

The third most often-referred-to medical document is the Kahun Gynecological Papyrus, which was discovered in 1889 and dates to 1825 B.C.E. As its name indicates, this document focuses on female health issues, including the description of a method for revealing whether or not a woman was pregnant and to determine the gender of the fetus. The process involved using wheat and barley and moistening both with a woman's urine. If the barley grew, she was pregnant with a boy; if the wheat grew, the baby was a girl. If nothing grew, she was not pregnant. (In modern medicine, the first reliable pregnancy test was not created until 1929. It, too, was a test that involved checking the urine.)

Other significant papyri included: the Berlin Papyrus (1300 B.C.E.) and the Hearst Papyrus (1500 B.C.E.), both of which were similar to the Ebers Papyrus, containing more about prescriptions as well as a lot of magic. The London Papyrus, dating to 1350 B.C.E., consisted mostly of magic. Interestingly, the earlier documents reflect a more rational approach to medicine, with magic increasing over time.

PREPARATION FOR THE AFTERLIFE LED TO EMBALMING

Preservation of the body was very important to the Egyptians; they had a deep belief in life after death and wanted to be certain that the person would have access to his or her body.

Before the Egyptians developed the concept of tombs and pyramids (as early as 5000 B.C.E.), Egyptians buried their dead in shallow graves in the desert. The hot, dry air proved perfect for drying out the body and created a natural mummification process. In about 3100 B.C.E., customs changed, and the bodies of kings and nobility were placed in beautiful, highly decorated tombs. To

Anubis is the Greek name for the jackal-headed god associated with mummification and the afterlife in Egyptian mythology.

the consternation of the Egyptians, the bodies thus honored rapidly decomposed, making them seem useless for the afterlife. In search of a solution, the Egyptians began to experiment with ways to preserve the body so that it could be reused in the afterlife.

By experimenting with naturally occurring chemical agents, the Egyptians created a successful mummification method by the Fourth Dynasty (2613–2494 B.C.E.); evidence of intentional mummification dates to that time. The key chemical used was natron, an indigenous deposit that occurs in dried lakebeds in certain Egyptian valleys. Natron is composed of sodium carbonate and sodium bicarbonate with sodium chloride and sodium sulphate, and it proved very effective at drying out a dead body and preserving the tissues. Because enzymes and microorganisms are unable to function without water, the drying out of the tissues prevented decomposition and served as a fixative.

Embalming Methods

Although the Greek historian Herodotus was writing a bit later, he and another Greek, Diodorus Siculus (first century B.C.E.) noted the methods of embalming that were used. The following three methods varied in thoroughness and price point; they are listed from most to least expensive:

1. The top-of-the-line method involved the removal of the internal organs. A hook was used to remove the brain through the nostrils. (The heart was considered the "seat of the mind," so it was often left in place.) Next, a cut along the side of the body with a sharp stone permitted the embalmer to slip the intestines out without disturbing the main part of the body. Muscles were removed through multiple small incisions. Once the abdominal organs were removed and stored in jars set aside for this purpose (canopic jars), the body cavity was cleansed with palm wine, and myrrh, cassia, and spices were added to provide fragrance. Sometimes, the organs were wrapped and replaced in the body. Next the body was covered in natron for 70 days, washed, and then wrapped in linen bandages, which were then smeared with gum to create a gluelike casing. The final step was to return the mummified body to the family, who would place it in a case and keep it in a burial chamber. If the organs had not been put back into the body, then the canopic jar was also given to the family to be stored along with the mummy.
2. A less costly method of mummification simplified the number of steps by filling the body cavity with cedar oil (injected through the rectum), which causes the inner organs to liquefy. The body was then covered with natron for 70 days. At the end of this period, the oil was permitted to drain out of the body, leaving only skin and bone.
3. The final method—and definitely the cheapest—was to wash the body in a salt solution, embalm it with natron for the 70-day period, and return it to the family for burial. Though it was certainly invasive, embalming did

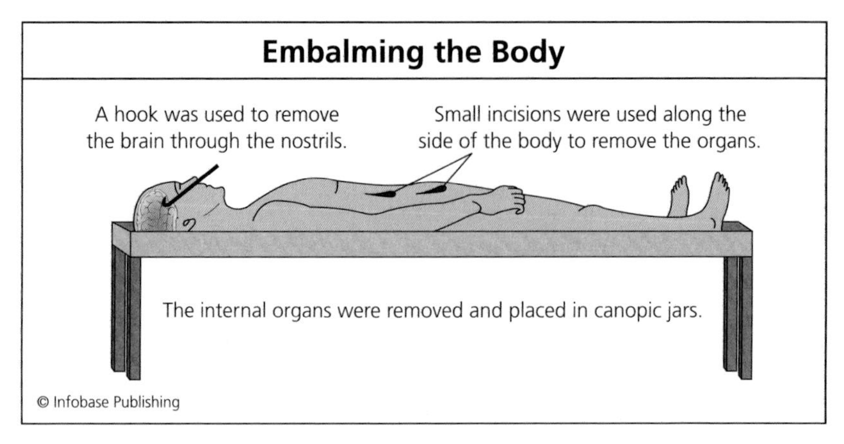
Embalming the Body
A hook was used to remove the brain through the nostrils.
Small incisions were used along the side of the body to remove the organs.
The internal organs were removed and placed in canopic jars.
© Infobase Publishing

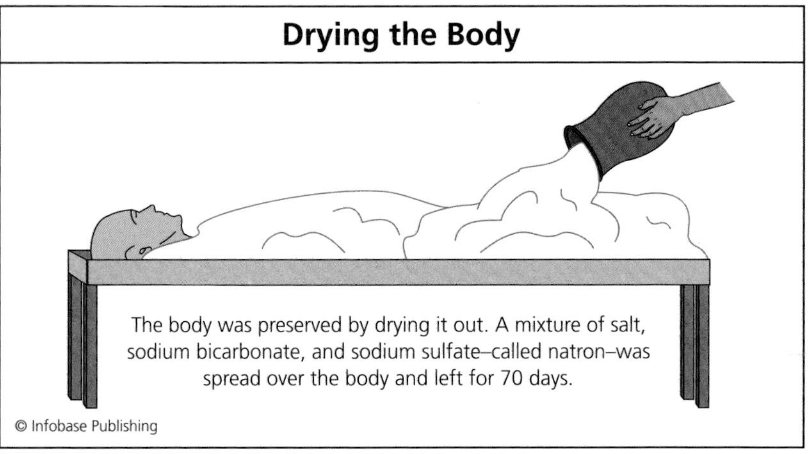
Drying the Body
The body was preserved by drying it out. A mixture of salt, sodium bicarbonate, and sodium sulfate–called natron–was spread over the body and left for 70 days.
© Infobase Publishing

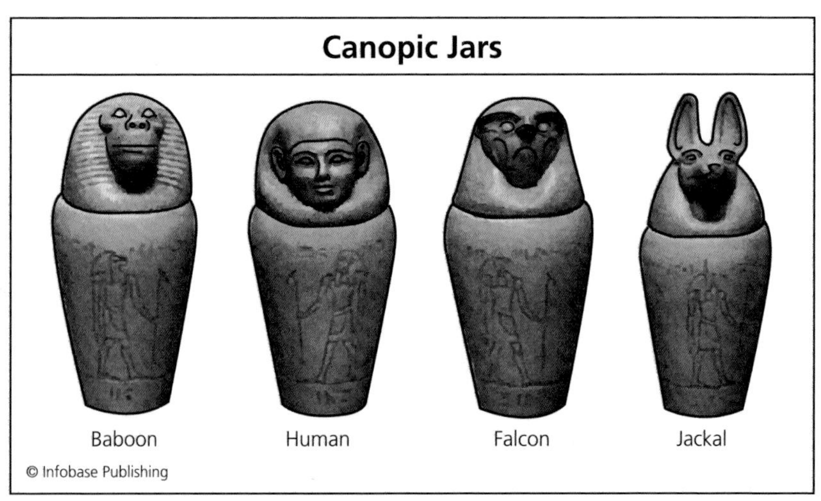
Canopic Jars
Baboon
Human
Falcon
Jackal
© Infobase Publishing

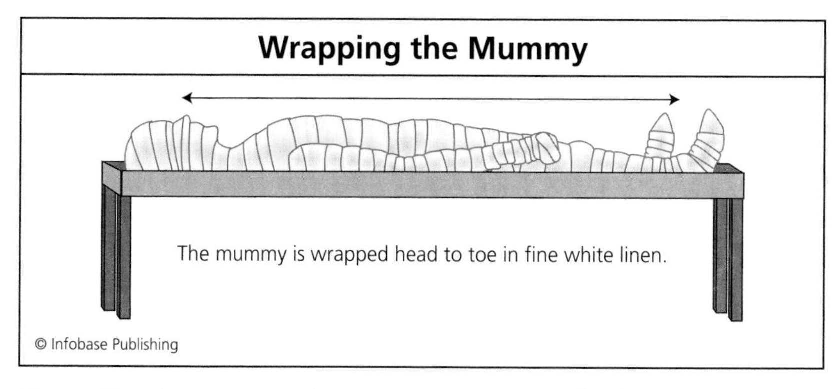

Wrapping the Mummy
The mummy is wrapped head to toe in fine white linen.
© Infobase Publishing

Mummification was an intricate process—available at a range of prices—that preserved the body for the deceased person's journey to the afterlife.

not lead the Egyptians to a better understanding of the body. Egyptian healers actually knew very little about human anatomy, and this had to do with the class or workers who did the embalming as well as the methods they used for organ removal. Because embalmers were of a lower social caste than healers, it was not considered appropriate for them to try to point out to healers anything they might have observed. In addition, the point of the embalming was preservation of the outer shell of the body, so the manner in which the organs were removed was less than ideal for studying anything—from function to placement.

Mummification was used until the seventh century C.E., surviving early Christianity but dying out with the beginning of Islamism.

Studying the Mummies

When doctors and scientists in the laboratory and professors in the classroom first began to study mummies, they would simply unwrap the linen wrap covering the body and examine it, a process that usually made the mummy unusable for study later on. Thomas Pettigrew, a 19th-century surgeon and antiquarian, turned this practice of mummy "unrollings" into a party entertainment where he would unroll the mummy and then perform an autopsy in front of his guests.

It was also common practice for 19th- and early 20th-century explorers (generally Europeans) to ship home their findings from the lands they explored. The removal of mummies from their very dry climate was not helpful in preserving them, and the legality of removing items from their native lands is only now being ironed out in courts around the world. Even when studied in Egypt, scientists frequently find that the mummy skin flakes or peels away because of being so old. In addition, sand preserved along with the mummy is also problematic. It needs to be cleaned away to study the body, but brushing it away can lead to more flaking of the skin.

Prosthetic limbs (primarily arms and feet) and a fake penis have been found with mummies. It is not clear whether these body parts were used in real life or provided to help people in the afterlife. The mummy of a 50- or 60-year-old woman who lived during the reign of Amenhotep II was found with a prosthetic toe. While it may have been made expressly for the afterlife, it seems that the wound had healed (indicating that the person lived), and it could have been very helpful for walking during her lifetime.

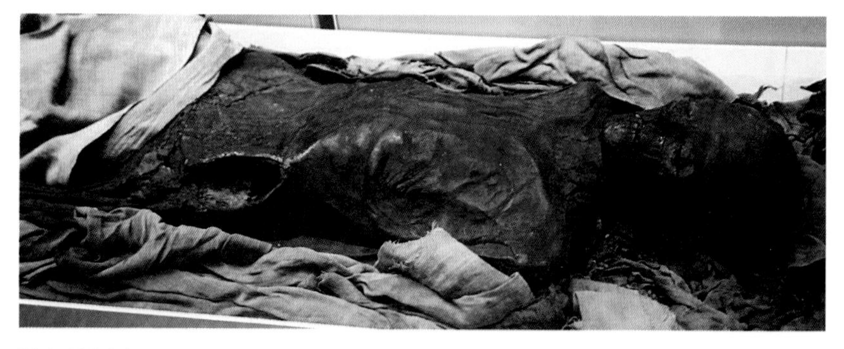

This British mummy shows how the body looks after it is unwrapped. *(David Monniaux)*

Only in the latter part of the 19th and early 20th century, with the birth of paleopathology, did the study of mummies become more scientific (see the sidebar on "Sir Marc Armand Ruffer"). In time, scientists developed less invasive ways to investigate these mummified remains. Scientists now use CT scans, three-dimensional imaging, electron microscopes, carbon dating, serological tests, DNA studies, fingerprinting, and dental studies as well as computerized facial reconstruction to study mummies. Recently, a team at the Manchester Museum in England pioneered the use of nondestructive techniques for mummy study, using such tools as endoscopy for viewing within the cavity or for snipping small bits of tissue. A tissue bank has now been established at Manchester University's School of Biological Sciences to store samples.

Egyptian Medicine

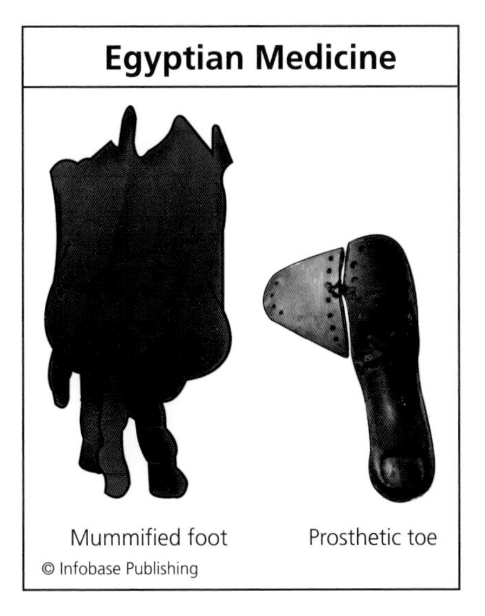

A few Egyptian mummies were found with prosthetic appendages such as this prosthetic toe. It is not known whether these were used by the people while alive or whether this was part of their preparation for the afterlife.

THE BUILDING OF THE PYRAMIDS AND THE HEALTH OF THE WORKERS

The pyramids, one of the most remarkable aspects of ancient Egyptian culture, were mammoth tombs for dead rulers; they provided space for the ruler's mummified body and all the material possessions that he or she would need in the afterlife. While archaeologists and paleopathologists have found the contents of the pyramids to be quite helpful in piecing together a picture of the Egyptians' attitudes toward health and their medical treatment, the act of building these colossal pyramids had a direct bearing on the medicine of ancient Egypt because of the types of injuries and illnesses brought on by the work conditions. There is ongoing discussion as to exactly how these mammoth structures were built (did they use levers to lift the stones? Were there ramps that could be used to drag the stone into position?), but the point on which there is agreement is that a vast number of men were required to build each pyramid and that many must have been injured during the process of moving such massive stones with very little mechanical equipment.

While there is sometimes a tendency to overmodernize interpretations of ancient practices, the Turin Papyrus actually reflects that the workforce benefited from a form of health "insurance." While there was no age limit or exemption from working on a project of the pharaoh's, there is mention of exemptions or assignment to "light duties" and even pensions for someone who was deemed incapacitated. The Papyrus of Anastasi IV shows that workers received the equivalent of pensions and sick leave.

There were two particularly challenging aspects of the work that would have affected the workers' health. First, the general work conditions involved extensive exposure to the sun and blowing sand. As a result, working hours were regulated to avoid sunstroke, and it was specified that workers were on duty for four hours in the morning and four hours in afternoon with a meal and a nap during the midday break. Second, the constant exposure to blowing sand meant that many Egyptians suffered from eye ailments, so the medical treatment for eye-related issues is more advanced than in many other cultures.

Broken bones were a frequent problem in working with these massive stones, but skeletal evidence shows that the Egyptians knew how to set bones. The Hearst Papyrus describes a plaster cast made of cow's milk, barley, and acacia leaves bound together with gum and water. The physicians may also have used splints and traction. Occasionally, a limb had to be amputated, but there are sign of healing at the bone ends, so people survived.

SIR MARC ARMAND RUFFER (1859–1917): *First Paleopathologist*

Scientists, doctors, and historians rarely make one giant discovery. As individuals, they each contribute small pieces of information to an overall process, and these small steps eventually lead to a bigger plan. Sir Marc Armand Ruffer (1859–1917) is generally credited as the first modern paleopathologist and the person responsible for defining the field itself in 1910. While his contributions were quite significant, he "stood on the shoulders of others" in his study of the Egyptian mummies.

Born into a well-known family in Lyons, France, Ruffer grew up in homes in France and Germany before moving to England, where he studied medicine. He went on to work at the Pasteur Institute (while Pasteur was still active there), eventually moving to the British Institute of Preventive Medicine. While studying diphtheria bacillus, Ruffer became quite sick from the disease and had to resign. He and his wife moved to Egypt for his convalescence.

Soon, he became a professor of bacteriology at Cairo, eventually working as president of the Sanitary, Maritime, and Quarantine Council of Egypt. One of his major accomplishments was eliminating cholera from the country by enforcing quarantine when necessary. At the outbreak of World War I, he worked as the head of the Red Cross in Egypt. He was knighted in 1916 but died at sea when returning from Greece, where he was helping reorganize their sanitary commission.

Ruffer was a well-respected bacteriologist and hygienist of international repute, but it was his contributions to paleopathology that make him memorable today. While others who had preceded Ruffer by 60 years or so (the 1850s) had found rudimentary ways to separate tissue samples from mummies so that tendons, cartilages, nerves, muscles, and fat

(continues)

(continued)

could all be studied, Ruffer developed a rehydration method that permitted the sectioning of the brittle tissues so that they could be stained and studied. (The rehydration restores the tissue architecture, but it is a delicate balance—too much hydration and the sample is ruined.) Ruffer's solution was a formulation of distilled water, ethanol, and aqueous sodium carbonate with an added fixative. In Ruffer's day, there was no way to document photographically what he saw under the microscope, so his findings were often drawn and tinted with color by his wife, Lady Ruffer.

Ruffer's observations were unprecedented for the day and identified a number of diseases from which ancient Egyptians had suffered. His identification of calcified eggs of *Schistosoma haematobium bilharzias* was the first indication that this parasite, which still plagues Egyptians, existed at this early date. Ruffer's studies also showed evidence of malaria, smallpox, and tuberculosis as well as signs of degenerative arterial disease.

The term *paleopathology* was coined by Ruffer to refer to the science of diseases that can be demonstrated in human and animal remains from ancient times. Today, the field relies on all types of testing from macroscopic methods (visual examination) to microscopic ones as well as immunologic, chemical, and mass spectrophotometry methods for studying skeletons, soft tissues, and even animal droppings (scat).

EGYPTIAN MEDICINES

While the Egyptians will long be remembered for their preservation of bodies through mummification, they also advanced the field of healing medicines. The Ebers Papyrus alone (and what exists is only a portion of this document) mentions 876 different prescriptions made from approximately 500 substances; each prescription usually involved many ingredients. Their quest for learning in this area was notable in that the documents sometimes specified what type of soil a particular curative plant needed to be grown in to be effective for a particular illness.

Medications were created as drinks, gargling solutions, poultices (a soft packing mixture to soothe or heal a wound), and snuff (something sniffed). The main liquids for medical mixtures were water, milk, wine, and the urine of an elephant cow.

Many of the basic ingredients in medicines were common fruits and vegetables that were viewed as medicinal. Extracts of plants (senna, colocynth, and castor oil) were sometimes used quite effectively as a way to purge (clean out) the system. Tannin was used for burn treatment. Tree resins (frankincense, myrrh, manna) and animal fat were also important, and minerals were used as astringents or antiseptics. Animal dung, lizard blood, and minute bits of fly carcasses were frequently cited as ingredients as well. (There is some thought that tetanus was actually caused by the use of animal dung on wounds.)

Like other cultures, a good number of the remedies in Egypt were selected based on "likeness" to the *symptoms* ("*simila similibus*"). For example, an ostrich egg was part of the remedy for a broken skull; a hedgehog, covered in spiny hairs, provided some of the ingredients for curing baldness, and so on.

Because the writing is in hieroglyphics, some of the ingredients are hard to translate and may or may not relate to substances known today. Did a reference to a "buttercup" mean a name of a plant or a "cup of butter (or its equivalent)"? When a medication specified "pig's teeth," did this mean teeth from real pigs or a substance that was simply called "pig's teeth"?

Many of the solutions the Egyptians found actually make good sense to doctors and pharmacists today. For example, raw liver, rich in vitamin A, was thought to cure night blindness, and night blindness is actually sometimes the result of too little vitamin A. The Ebers Papyrus recommends the creation of a vaginal contraceptive solution made of crocodile oil, gum acacia or honey, and natron (the natural chemical also used in embalming). When dissolved, the gum acacia forms lactic acid, and this would have served as an effective spermicide. They also knew about opium poppy, hellebore, henbane, and mandrake, all of which could have been quite powerful. (For more information on remedies used, see the accompanying chart showing "Common Egyptian Cures.")

The Greek historian Herodotus (ca. 484–between 430 and 420 B.C.E.) writes that during the building of the pyramids, workers were given enormous amounts of radish, garlic, and onion. This is interesting in light of 20th-century discovery of antibiotics that have been extracted from radish. Allicin is an antibacterial and antifungal compound from garlic and onion. There is some documentation of the use of yeast, which was discovered to be a helpful antibiotic in the early 20th century.

EARLY DENTISTRY

Good teeth have always been a key factor in overall good health. Ancient cultures appreciated the importance of this since teeth were likely to have been used as tools of sorts to crack open or loosen something. In addition, strong teeth would have been necessary for chewing food that was less refined than what exists today.

Based on the information in various medical papyri as well as evidence from skeletal remains, abrasion was the cause of a great deal of tooth damage suffered by Egyptians. The blowing sand and grit may have mixed into the food and caused enough wearing away of the tooth enamel that infection and resulting bone destruction often occurred. In addition, the decay in Egyptian teeth is generally found around the gum margin—this type of dental decay was likely caused by the type of bread they ate. Made from emmer wheat, it was particularly sticky and may have stuck to the tooth base, causing tooth decay.

For most Egyptians, dental issues were usually solved by their doctor, the swnw, but for Egyptians of rank, there were two classes of dentist: dentist and great dentist. (Reference to these specialties dates to 2650 B.C.E.) As to the type of dental care administered, the

COMMON EGYPTIAN CURES

Note: The following are a few examples of cures that were used by the ancient Egyptians.

HERBAL REMEDY	USE
Coriander	for digestive ailments due to its cooling and stimulating properties
Cumin	mixed with wheat flour and water to form a paste and then applied to arthritic joints
Honey	antibacterial and antibiotic properties; used in treatment of open wounds
Mint	for gastric disorders and fresher breath!
Oil of fir	natural antiseptic—helpful with wounds
Pomegranate	contains a natural alkaloid (tannin) base that is effective in paralyzing a worm's nervous system. When infused with water and drunk it would have cleared the body of roundworms and tape worms.
Poppy	could be broken down to produce a liquid that had extensive narcotic properties and could be used for pain relief. It was also added to water to be used as a sleeping pill.
Willow	could be crushed to form a paste and used to relieve a toothache. Also used to treat burns when combined with sycamore and acacia.

Egyptians had several solutions. One translation refers to chewing on willow. A word is missing, so the information is incomplete, but willow eventually was used to create aspirin, so there was basic wisdom in this prescription. In other cases, a throbbing tooth very likely was pulled.

Several remedies for bad breath (halitosis) are noted in various medical papyri so it was a mouth-related issue of which Egyptians were highly aware. One regimen involved using natron pellets (the substance used in embalming) as a chewing gum; another description of breath sweetener involved this: "Take frankincense, myrrh, cinnamon, bark, and other fragrant plants, boil with honey, and shape into pellets."

CONCLUSION

Egyptians made many important medical advances, and they understood the workings of a variety of medicinal drugs as well as some contraceptive devices. During the time when ancient Egyptian culture thrived, many of their ideas on health and medicine were passed on to the Greeks, who then passed the knowledge on to future generations. In addition, the process of mummification and the development of writing has permitted information about the Egyptians and their practices to become known to scholars and scientists of a much later period.

But Egypt was not the only country making rapid strides forward. India, too, was developing ways for communities to handle sanitation, and the Indian culture excelled at surgical techniques that will be discussed in the next chapter.

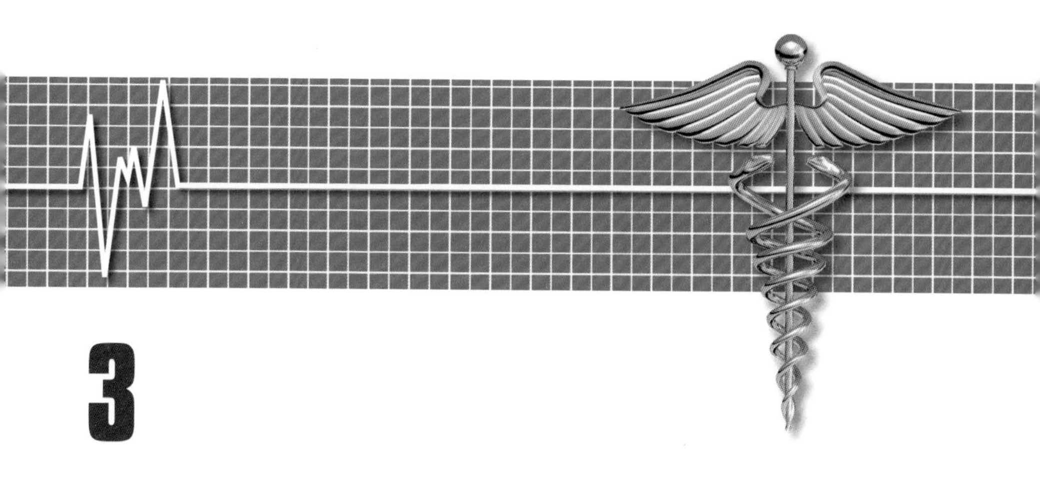

3

Indian Influence on Medicine

For many years, the history of medicine almost totally ignored the medical practices of early India and China because of the manner in which the two countries developed. During ancient times, these two countries operated as isolated entities that underwent a high number of disruptions of their governing bodies. While active religious and medical traditions were very much a part of both cultures, fragmentation of the area by warlords and other interferences prevented an orderly flow of information to other parts of the world. The languages also presented barriers. The Sanskrit writing used in India as well as the hard-to-decipher characters of the Chinese language meant that communication about their medical practices and their considerable achievements was difficult.

Despite late recognition, some of the medical practices of both ancient India and China have proven to be worthy of interest to practitioners of today. Indian medical practitioners put forward a belief in a life of moderation that can lead to natural healing, and these are considered important components of alternative medicine practices of the 21st century. (The Chinese history of medicine and its modern influence will be explored in the following chapter.)

Early Indian medicine may have begun in an advanced civilization in northwest India along the Indus River Valley, and this chapter will begin with what scientists know of these communities and their practice of medicine. Like their neighbors in Egypt and Mesopotamia, these people seemed to have developed a remarkable culture before their civilization all but disappeared in approximately 1500 B.C.E.

The history of Indian medicine in the years that followed can be divided into two periods: the Vedic period (until about 800 B.C.E) when the medical information derived primarily from the *Vedas,* four holy books written in Sanskrit, and the period of the Golden Age of Indian Medicine (800 B.C.E.–1000 C.E.) when Ayurvedic medicine blossomed. As this chapter will show, India's medical progress was very much entwined with the religious development in the country, which was first influenced by the growth of the Hindu religion and soon followed by the development of Buddhism.

ANCIENT CIVILIZATIONS IN INDIA: HEALTH AND MEDICAL ADVANCES

Like the early settlements in Mesopotamia and Egypt, ancient civilizations in India developed along major rivers. While evidence of medical progress from this early time is far from abundant, various clues within the community indicate sophistication in the practice of medicine that would match the more advanced settlements along the Nile, the Euphrates, and the Tigris Rivers.

Little was known of these early settlements in India until the 1920s, when some engraved seals bearing pictographic script were found near present-day Saniwal, in the Punjab region of what is now Pakistan. This discovery led to the eventual uncovering of almost 100 communities in northwestern India in the Indus River Valley. Now referred to as the Harappan civilization, this society existed from approximately 3000 B.C.E. until about 1500 B.C.E.

The two major cities of this civilization were Harappa, located near what is now the Punjab region, and Mohenjo-daro ("mound

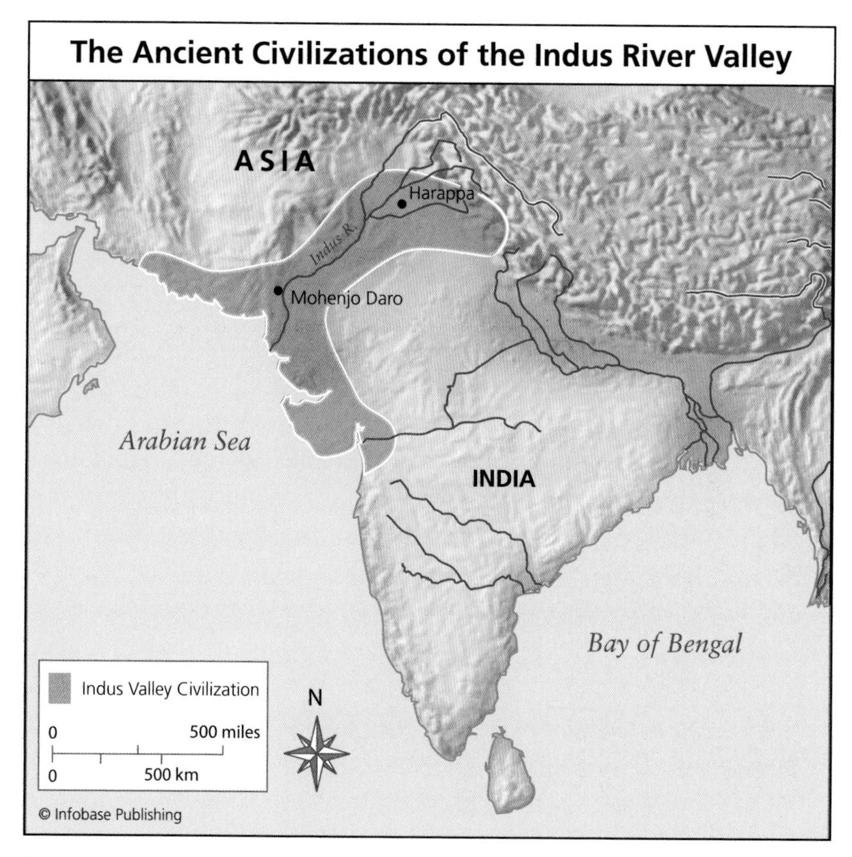

The Indus River Valley was the location of an advanced ancient civilization in India. Little was known about it until a discovery in the 1920s.

of the dead"), which is more inland but still closer to the mouth of the river; these cities are thought to have been the seats of government. Parallel in design, the cities were built according to a grid pattern with broad streets that divided the towns into compact regular blocks. In the citadel area, archaeologists have identified a granary, two large assembly halls, a huge brick-lined water tank, and a single residential structure. The residential neighborhood in both cities was built east of the citadel, and the streets were laid out in a grid pattern.

The importance of hygiene is clear from the community provisions for city sanitation, and these measures would have been a good first step in reducing exposure to some illnesses. Rubbish was disposed of in outlying areas, and many homes featured water that was piped inside. The cities also had extensive drainage systems to carry away wastewater; the drainpipes featured manhole covers that must have permitted pipes to be inspected or cleaned. This level of sanitation was equal to what Rome would eventually achieve 1,000 years later.

The Harappan agricultural community developed irrigation methods as well as basic methods of flood control, and they were early in the domestication of animals including cattle, dogs, and a few elephants. (The civilization is thought to have been the first to harness elephant power to help with agricultural work.) However, the maintenance of cattle herds and other animals brought with it exposure to pathogens, and there is evidence of

YOGA AND ITS EARLY ORIGINS

Yoga is a program of exercises for physical and mental nourishment that came about in ancient India. The practice of yoga involves a complex series of postures that are combined with conscious use of breathing and meditation to calm the mind and establish harmony between mind and body. In yoga, the mind, the spirit, and the body are considered one, and it is believed that if a person can bring about unity within oneself, then this unity will create self-healing.

The earliest known origins of yoga date to the Harappan civilization. Yoga positions were depicted on some of the seals found near Mohenjo-daro. One shows a figure standing on its head, and another shows a person sitting cross-legged in what are both classic poses of yoga.

Though the Harappan civilization faded away, the practice of yoga continued to appeal to the Hindu people who came to dominate the Indus River Valley. As yoga evolved, some aspects of the practice of yoga derived from the Vedic reli-

smallpox in this area as early as 3000 B.C.E. This is further supported by signs that there were temples that may have been built to worship a smallpox deity. (Each disease was associated with a different god.) In addition, India was one among several ancient places, including China and Persia, where healers were thought to have used a scarification process to inject a small amount of smallpox virus and thus inoculate people against the disease. (In the West, it would take until the 18th century before an English physician was able to prove the value of this type of protection against disease.)

Archaeologists can identify more than 400 distinct symbols in the pictograph script that was on the seals found in the 1920s, but translation has been slow. It is unclear whether this was actually a form of writing or whether it was a nonlinguistic signage system

gious texts, which are the foundation of Indian Hinduism. The Hindus believed that good health was built around a lifestyle of control and self-discipline; the term *yoga,* from the Sanskrit word *yoktra,* means yoke. This is generally interpreted to mean "to yoke or harness the mind" since concentration is a key element in yoga.

One of the Hindu scholars, Patañjali, who lived in the second century B.C.E. was at one time thought to have been the "father of yoga." Now it is known that yoga predated him, but he collected much of the writing about yoga and is thought to have written at least part of the *Yoga sūtras.* According to Patañjali's writings, the human body contains channels that connect seven centers of spiritual energy, known as "chakras." If an individual learns to focus and control these "chakras," then the hidden energy within the body (called Kundalini) can be released. The release of it permits the body to do more than it would normally be capable of accomplishing.

(marking of things). Even if it was only rudimentary markings, knowledge has still been gained from the pictograph script. For example, historians know that the practice of *yoga* dates to this time (if not earlier) as pictographic script from this time shows at least two figures in classic yoga poses. (For more information, see the sidebar "Yoga and its Early Origins.")

Because of the difficulty of teasing out much information from the scant amount of writing available, the best clues about these communities are archaeological. Scientists, led by the anthropologist Professor Andrea Cucina from the University of Missouri-Columbia, have discovered teeth from this area that indicate that as early as 7,500 years ago (certainly by the time of the Harappan civilization), these Indus Valley inhabitants had knowledge of dentistry. Analysis of teeth found from this period shows that the Harappans knew about drilling through tooth enamel to remove areas of rotting tissue in what must have been an effort to ease tooth pain. At one excavation, scientists located a total of 11 teeth that had been hollowed out in this fashion. The scientists also found flint drill heads in abundance. Based on their high level of advancement in the field of oral health, they would likely have developed skills for managing other types of bodily ills as well.

The height of this civilization seems to have been about 2500 B.C.E.; 1,000 years later, this civilization had all but disappeared. While there is speculation that invaders may have hastened the end, most experts think that changes in the river led to a depletion in the topsoil, so the people moved on to look for more fertile land. The progress in sanitation, agriculture, and what appeared to be peaceful government (the cities showed no signs of needing to protect against invaders), as well as the pictographic marking system, all vanished with the disappearance of the civilization.

VEDIC MEDICINE

As the civilization around the Indus River declined, Indo-Europeans migrated into the river valley. They may have destroyed any remaining members of the previous Harappan civilization, and those who settled in the area were to become the early followers of Hindu. Vedic medicine began as an oral tradition that was passed down from teacher to student. The basic belief of this early time was that actions determined destiny and that sin was the cause of disease. As a result, the importance of leading a good and moral life was an early focus of Vedic medicine.

Good health was built on a platform of careful hygiene, good diet, and exercise. The Hindus believed in regular bathing, used special oils to anoint the skin, and advocated tooth cleaning, which was often accomplished by chewing on betel leaves (a plant that was used for many medicinal purposes, including the freshening of breath). When it came to diet specifications, water was one of the cornerstones of good health, and Hindus went so far as to specify the amount of water to be consumed before and after a meal. Food safety was part of a proper diet, and they knew not to eat fly-infested foods. As part of cleansing their digestive organs,

The Hindu Trinity—Brahma, Siva, and Vishnu—Hoysaleswara temple, Halebid *(Calvin Krishy)*

Hindus used one of five sanctioned procedures: emetics, purgatives, water enemas, oil enemas, and sneezing powders.

The methods for healing in Vedic medicine were heavily reliant on religious beliefs with a good dose of magic. Cures often involved confession of the sin as well as exorcisms and spells. Certain gods were associated with certain illnesses, and if someone was ill, the evil spirits were drawn out by mantra (incantations), supplications, and expiation. Bleeding (often with leeches) was also an important aspect of cures at this time.

Vedic practices were eventually documented in writing. The *Vedas* (the body of sacred Hindu writings, chief among which are four books, the Rig-Veda, the Sama-Veda, the Atharva-Veda, and the Yajur-Veda) conveyed an entire religious and intellectual *philosophy* of life but also included information about treatments for wounds, snakebites, and the removal of arrows as well as the amputation of limbs. This written form has provided the term *Vedic medicine.* Because much of the material was written down much later than when the practices actually began, the cures are sometimes the "ideal" rather than what actually was practiced.

The early Hindu people had a simplified caste system that eventually became much stricter and more stratified. As the lines between castes became more firmly drawn and more stratified, it had a definite effect on medicine. Early on, the caste system included *Brahmans* (priests, including healers); *Kshatriyas* (kings and soldiers); *Vaisyas* (peasants and traders); *Sudras* (lower caste workers who were still considered "acceptable"); *Pariahs,* (the darker-skinned outcasts who eventually became the group known as the "untouchables.") The caste system grew to include at least 100 castes and subcastes. Later, this caste system led to a slowing of medical progress and treatment. The Hindus came to believe that it was taboo to touch another person, a belief that was primarily observed by the upper castes who were then limited to health checks that involved checking the pulse and examining the tongue. Lower-caste members did not observe this as strictly so they continued to explore some surgical cures, but the field was definitely slowed by this change.

Dhanvantari, god of Ayurveda *(Dhanvantari Ayurveda Center)*

THE ORIGINS OF AYURVEDIC MEDICINE

Ayurveda (from *ayur,* meaning life, and *veda,* meaning knowledge) grew from the Vedic tradition that preceded it. Yet the medicine that developed dropped many of the magical elements and became more empirical. The primary objective continued to be the maintenance of health, not the treatment of disease. When disease occurred, Ayurvedic medicine attempted to focus on doing away with the cause of the illness, not just getting rid of the symptoms. Ayurvedic practitioners believed there were four pillars of healing (the physician, the attendant, the medicine, and the patient), all of whom had to be involved in the process.

Ayurvedic physiology relied on a concept of balance of the "dosas" (like the Greek humors) in the body. The dosas of the body included wind, bile, and phlegm, and these were thought to align with the powers of nature: the wind, the Sun, and the Moon. When people suffered an imbalance of these dosas, it could lead to a disturbance of the blood, so the physician often had to remove "bad blood" by venesection (cutting of *vein*s) or leeching. Further balance was achieved by proper diet, which in Indian lore was a vital element in maintaining health. (Meat-eating was still permitted at this early time, and the reverential treatment of animals did not come about until the sixth century C.E.)

Ayurvedic medicine, like the writings of Hippocrates, was not documented by a single individual. The teachings were compiled in several books by a few contributors. The primary documents that survive include those assembled by Caraka and Sushruta. Both healers shared a common intellectual tradition, and their writings created the cornerstone of Ayurveda. Caraka's writings contain long reflective passages followed by descriptions of many illnesses and the recommended cures. Sushruta's work was more heavily focused on surgical methods; this was an area in which ancient Indian practitioners excelled.

Caraka-samhita: Early Medical Writings

The Caraka-samhita was the name given to the collected writings of Caraka, a healer who is thought to have lived some time between

1000 and 800 B.C.E. The medical methods noted by Caraka had developed long before his time, but Caraka noted them and refined them. His work is thought to have been the first treatise of Indian medicine, and it was so influential that it became a standard textbook that was used for almost 2,000 years.

The Caraka-samhita identified 107 vital points (*marmas*) by which illnesses and injuries were referred. *Marma* were points at major veins, arteries, ligaments, joints, and muscles where injuries were likely to be fatal. Diagnosis of specific illnesses was daunting because even at this early time, more than 1,000 diseases were described by Caraka.

To arrive at a diagnosis, the physician listened closely to the person's description of the illness. Then he studied general appearance noting any abnormalities. An examination was based on touch (palpation) and auscultation (listening to internal noises). After this, he examined the blood, body fluids, and excretions. If a physician did not want to taste the excretions himself, he could assign it to his students or feed the excretions to insects and observe reactions. They used a diagnostic taste test for what they referred to as "honey urine disease," probably what we know as diabetes.

The Caraka-samhita outlines an oath very similar to the *Hippocratic Oath* (that was to come later, in approximately the fifth century B.C.E.). Ayurvedic practitioners had to vow to be celibate, speak the truth, eat vegetarian, never carry weapons, obey the master, and pledge devotion to his patients (including not to have sex with them). Practitioners also were forbidden to treat enemies of the king and could not treat women unless they were accompanied by their husbands or guardians. Students who were not yet fully trained had to be chaperoned when seeing patients.

Sushruta-samhita

The Caraka-samhita was followed by the Sushruta-samhita, the oldest treatise about surgery. It is thought to have been written about the same time as the height of Greek medicine (fifth century B.C.E.). Written by a well-respected medical practitioner by

Sushruta performing plastic surgery on the ear *(Central Council for Research in Ayurveda & Siddha, New Delhi, India)*

the name of Sushruta, this work was a compilation of the various types of surgery being performed in India in ancient times.

In his book, Sushruta details surgery under six categories: excision (Chedya), scarification (Lekhya), puncturing (Vedhya), exploration (Esya), extraction (Ahrya), suturing (Sivya). The provisions for surgery were detailed. There were special surgical beds, including one particularly for setting fractures. Twenty sharp implements and 101 blunt ones are mentioned among the surgical tools that were used. Forceps in various sizes (from "lion-mouthed and crocodile-mouthed to heron-mouthed") are described as are the types of needles and threads that should be used in sewing up surgical incisions. (For an innovative "stitching" method involving ants, see the next section.) Hot oils and tar were recommended for stopping bleeding.

Antiseptic care and anesthetic efforts were made even at this early time, likely out of poor experiences in ignoring matters of cleanliness or pain. The cultural emphasis on hygiene extended to the operating table, and infection may have been reduced as a result. The Sushruta-samhita describes methods of fumigating both the wound and the sickroom. Both Caraka and Sushruta refer to the drinking of wine before surgery, and one text refers to burning hemp (marijuana), which might have released a narcotic fume. But texts also refer to tying down the patient before an operation, so it seems obvious that the anesthetic system was imperfect at best.

To obtain proficiency in surgical skills, Sushruta devised various experimental methods to learn new techniques. For example, incisions and excisions were practiced on vegetables and leather bags filled with mud of different densities. The art of scraping was practiced on the hairy hides of animals; the puncturing of veins was tested on dead animals and lotus stalks, and the proper method for scarification was tested on wooden planks smeared with beeswax.

The Sushruta stipulated certain rules of practice. Pregnancies were usually managed by a *midwife,* but surgeons could "turn, flex, mutilate, or destroy" the baby if necessary. If the mother died during childbirth, the Sushruta recommended cesarean section. In general, physicians were given some latitude as to whom to treat. They were advised to be kind to all, but they were given permission to detach from those whom they determined were destined to die.

Sushruta was one of the first to study human anatomy, and in his writings, he describes how this can be accomplished without cutting into the body, a practice that was forbidden by the Hindus. Sushruta noted that for the purposes of study, the practitioner should select a human cadaver that is not too old or too young or one that has been overly ravaged by a serious disease. Once the body is selected, the intestines should be flushed out. Next, the body should be wrapped in grass or hemp and placed in a cage to protect it against animals, and then the cage should be dropped into a river where the water flow was no stronger than a gentle current. Seven days later, the cage can be lifted from the water. By brushing the layer of grass away slowly, the "eye can observe both the outer and inner part of the body as the skin will peel away with the brushing." This promised to give a better understanding of the bones, muscles, ligaments, and joint, but nerves, blood vessels and internal organs would have been difficult to study using this process. Despite the specificity of the description, it is not known whether or not this method was common practice.

SURGICAL ADVANCES IN INDIA

Surgical progress in India showed innovation as well as comprehension of what needed to be done. The surgeons performed excision of tumors, incision and draining of abscesses, perfected puncture methods to release the fluid in the abdomen, learned how to extract foreign bodies, mastered the repair of anal fistulas, and knew how to splint a broken bone. But there were a few areas where they truly excelled and accomplished successes that were unheard of until a much later time. One of the first surgeries at which they were superior was at the removal of bladder stones (lithotomy), developing a process that was not available to Europeans for another 2,000 years.

They also learned to repair torn intestines, perhaps partly because of an innovative "stitching" method they devised for closing interior wounds. The text described a method using large Bengali black ants after surgery to bite the edges of a wound. Once the ant had latched onto the edges of the wound, the head of the insect could be removed without disturbing the mandibles, which remained attached and kept the wound clamped closed.

Plastic surgery also reached remarkable heights with Indian surgeons. The Sushruta-samhita outlines some of the basic methods of this type of surgery, ranging from how to cover small defects to the rotation of skin flaps to cover up a loss of skin from a certain area. The work that was done ranged from simple repairs of the ear lobes, which frequently tore because of the custom of wearing heavy earrings. But there were other needs as well. In India, the punishment of a disloyal spouse was to cut off his or her nose. The cosmetic procedure of "nose remodeling" (rhinoplasty) became an important one for Ayurvedic healers to master. While there was

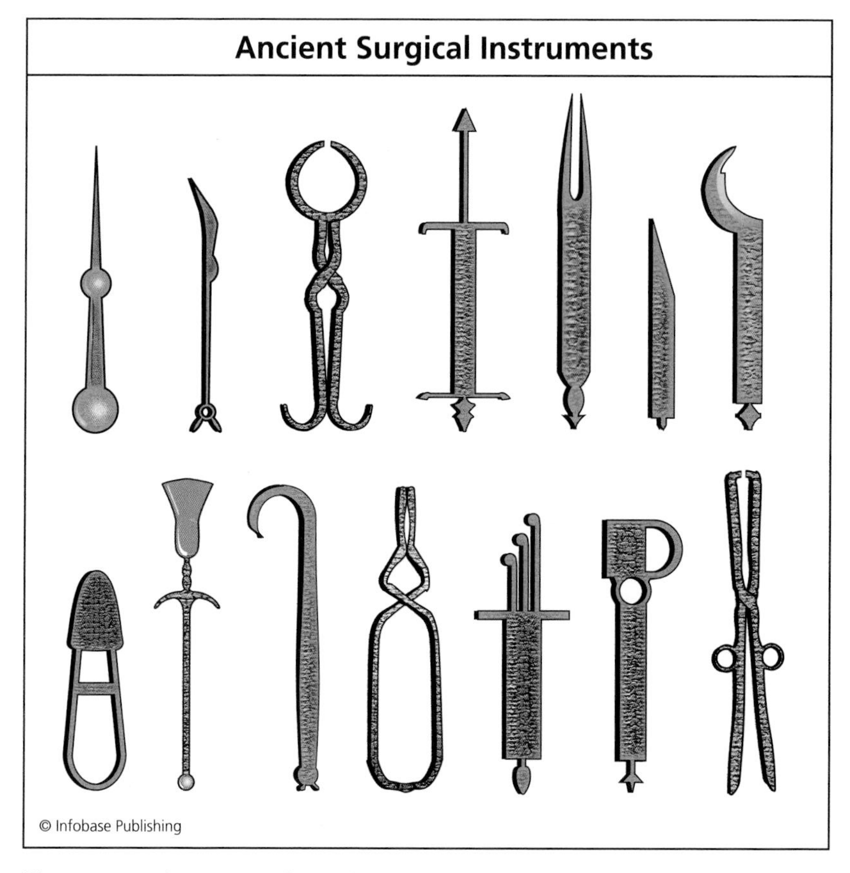

These were the types of surgical instruments used in India.

no direct description of the process in the Sushruta-samhita, the basic surgical methodologies were outlined, and the actual method that was used was clarified in the 19th century, when it was performed on a British bullock (a castrated bull) driver who lost the tip of his nose. The driver, accompanied by two British officers, happened on a brick maker near Poona who offered to make the nose repair, which he proceeded to do quite successfully. Later, the British officers, who also happened to be surgeons, wrote about the process for *Gentleman's Magazine* in 1794, calling it the "Hindu method." The system involved cutting a leaf-shaped flap of skin from the forehead, being certain that the section right by the nose

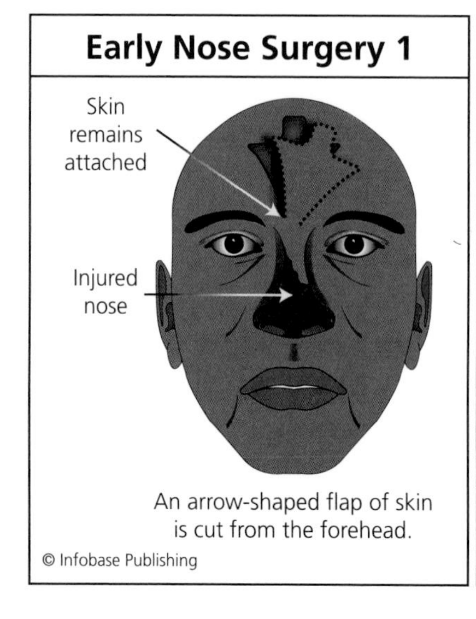
Early Nose Surgery 1
Skin remains attached
Injured nose
An arrow-shaped flap of skin is cut from the forehead.
© Infobase Publishing

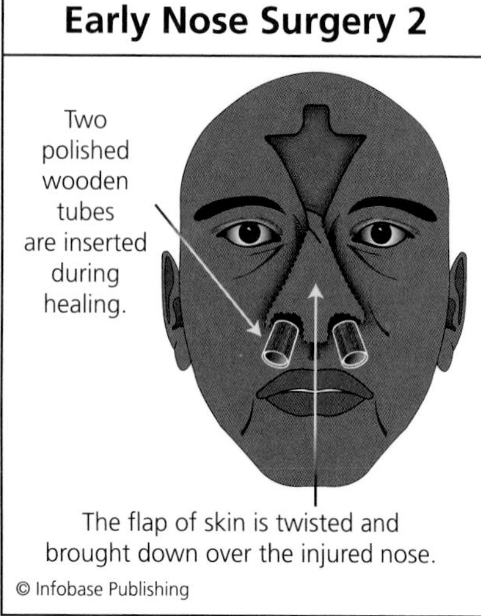
Early Nose Surgery 2
Two polished wooden tubes are inserted during healing.
The flap of skin is twisted and brought down over the injured nose.
© Infobase Publishing

was still attached. This loose "flap" of skin was then brought over the nose and stitched lightly so it could heal. When the skin flap had begun to grow in the new location over the nose successfully, the connection to the forehead would be severed. The surgeons reported that it was superior to any procedure that they had ever seen. (Because the caste system had slowed the progress of surgical methods being used by the upper castes, this may be the reason why a brick maker not only knew the procedure but also was the one willing to carry it out.)

Ancient Indian healers also are thought to have performed cataract surgery. The removal of cataracts was one of Sushruta's specialties, and the process used was this: First, the patient was fed and bathed and tied up in such a way that he or she would be unable to interrupt the procedure. The patient was then seated on the ground facing a knee-high bench on which the physician sat. The patient's head was held firmly. Next, the doctor warmed the patient's eye with his breath and asked the patient to look straight ahead at the doctor's knees. The doctor held the lancet between his forefinger, middle finger, and thumb and introduced it into the patient's eye toward the pupil, "half a finger's breadth from the

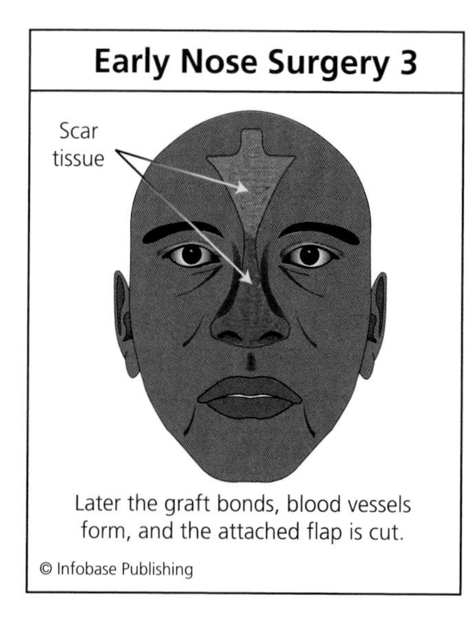

Early Nose Surgery 3
Scar
tissue
Later the graft bonds, blood vessels
form, and the attached flap is cut.
© Infobase Publishing

Cutting off the nose of someone who behaved badly was a frequently used punishment in ancient India. As a result, physicians mastered the art of nose surgery.

black of the eye and a quarter of a finger's breadth from the outer corner of the eye." He moved the lancet gracefully back and forth and upward. He waited for a small sound and for a drop of water to come out. Speaking a few words of comfort to the patient, the physician next moistened the eye with milk and then scratched the pupil lightly with the tip or the lancet to force the "slime" toward the nose. The patient was then told to inhale through the nose to get rid of the slime.

Before covering the eye with cotton soaked in animal fat and asking the patient to lie still for a time, the doctor "permitted the patient a few moments of joy" to experience seeing objects through the eye on which he had operated.

INFLUENCE OF BUDDHISM ON MEDICAL THINKING

Buddhism began in India in approximately the sixth century B.C.E., and as with other cultures of this time period, the religious beliefs of the people influenced the way that medicine was practiced. Its popularity early on made it the first world religion that spread beyond the society where it was born. Buddhism's early influence was on the health and the lifestyle of the people. Later, the pacifist beliefs led to a reduction in surgery that affected the medical care the people of ancient India received.

Buddhism was introduced by a prince, Siddhārtha Gautama (ca. 563–ca. 483 B.C.E.), who wanted to banish the strict caste system that dominated Hindu society and the religious rituals that controlled the Brahmanic priests. He renounced his own lofty position (that of the warrior class) to seek spiritual fulfillment. After much searching and experimenting with the ascetic way of life, he felt that he achieved enlightenment after meditating under a banyan tree at Buddha Gayā. He became known as the Buddha ("enlightened one") and traveled throughout the Ganges Valley, teaching the "four noble truths" of Buddhism. Buddhism preached universal love, service, the banishment of suffering, and the gain of peace of mind.

After the original Buddha's death (483 B.C.E.), however, the religion absorbed some of the preexisting beliefs of Hindu, so Buddhism mutated to be a different system that came to coexist with Hindu. A bit later, the religion received a big boost from a ruler, Aśoka (273–232 B.C.E.), a very fierce ruler who successfully united all of India but who became repulsed by his own soldiers' brutal methods. Remorseful about the type of leader he had been, Aśoka converted to Buddhism. During the remainder of his rule (261–ca. 232 B.C.E.), he made Buddhism the state religion. He devoted himself to righteousness and care of people and animals and used royal resources to provide community improvements such as building water reservoirs and wells. He also had shade trees planted so that weary travelers had protection from the strong sun. Rest houses also were built for travelers, and, later, these were converted to hospitals.

When Aśoka asked for an end to bloodshed, this was interpreted as reducing bloodshed of all sorts. Surgery became less common as a result, and the country reverted to the main tenets of Ayurvedic medicine—from moderation in diet and drink to a regular regimen of exercise—minus the surgery.

THE USE OF HERBAL MEDICINES

Herbal cures were a very important part of the Ayurvedic tradition. The people of ancient India identified and relied on the use of a large number of medicinal plants, concocting them from wide range of herbs. The Caraka-samhita cites 500 medicinal plants, while the Sushruta-samhita cites 760 various remedies ranging from milk of certain animals, minerals (sulfur, arsenic, lead, copper, sulfate, gold), and drugs made from various plants. They often had a high content of metals, particularly lead, mercury, and arsenic. The main liquids in which herbs were mixed for drinks were water, milk, wine, or elephant cow urine.

Ayurvedic medicine was notable for its belief in the slow and steady cure rather than the instant cures promised by modern-day antibiotics and painkillers. The herbal remedies used in Ayurvedic remedies are of a type that registers gradually within the body, and in the process, they set off fewer side effects. Among them are a good number that make sense to pharmacists today. Senna was used then and still is used as a laxative, and a compound from the Guggul tree has been shown in small studies to be helpful in controlling cholesterol levels.

CONCLUSION

By the 11th century C.E., the Muslim invasion of India brought Arabic medicine (Yunani medicine) into the country, and in time, the culture began to be influenced by the medical traditions of the Greeks and of Galen. Both Yunani medicine and Ayurvedic medicine are still practiced in parts of India today, with an overlay of medical care featuring some of the gains brought about by Western science.

Like the medical developments in India, the Chinese medical explorations took place in isolation and also proved to be quite effective. The next chapter will explore how Chinese medicine has come to influence Western medicine of today.

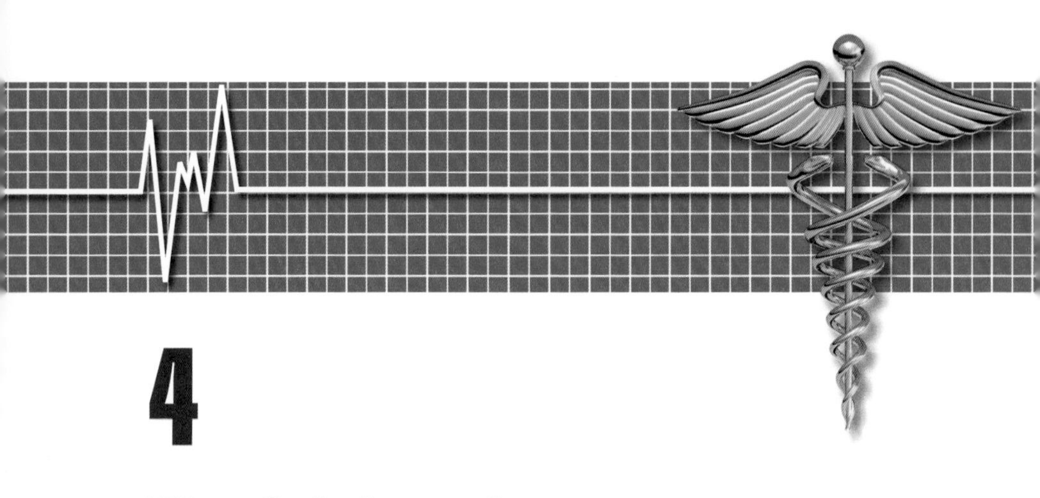

4

The Origins of Chinese Medicine

Traditional Chinese medicine is one of the world's oldest forms of medicine, yet until the last 40 years, Western practitioners knew little about it. A change occurred in 1971 when Secretary of State Henry Kissinger made a trip to China in advance of what would be President Richard Nixon's historic visit in 1972. James Reston, a reporter for the *New York Times,* was part of the press entourage accompanying Kissinger, and Reston became ill and required an appendectomy. The surgery was performed using *acupuncture* instead of anesthesia. To the amazement of the U.S. visitors, it proved to be quite effective, and this so impressed Kissinger that he encouraged President Nixon to implement a cultural exchange of medical practitioners between the United States and China.

Many of the beliefs that were put forward by the ancient Chinese have survived and are practiced in parts of the world today as holistic or integrative medicine. As in other ancient cultures, supernatural beliefs and religion were very important in the early practice of medicine, but in time, the Chinese people began to turn away from shamanlike healers as secular practitioners introduced a more naturalistic path to good health.

Even from the earliest times, the specific details of a disease have not been the focus of Chinese medicine. Instead, Chinese physicians consider the patient as a whole and prefer to work slowly toward an overall bodily cure rather than seeking a "quick fix" and treating a disease in isolation. It is sometimes said that Western practitioners treat a disease while Eastern practitioners treat the whole person.

The ancient Chinese people made advances in many fields including pharmacology, but for the Chinese, drug cures were a secondary choice. "A superb physician 'treats' before the ailment occurs" is attributed to the early and important Chinese work, the *Neijing* (*Nei Ching*), or *The Inner Canon of the Yellow Emperor.* For the Chinese, the first step in good health was conditioning the body. The Chinese created the art of Daoyin (Tao Yin, known today as Qigong—breathing exercises) and *taiji* (also known as tai chi ch'uan), a form of internal martial arts now thought to improve health and longevity. The Chinese also believed in the importance of proper eating to maintain health, and in the *Neijing* (*Nei Ching,* the document attributed to the Yellow Emperor) prescribes diet based on balance of the five elements.

The Chinese people valued diet and exercise, which they found resulted in improved health. When someone was ill, medical practitioners worked to bring balance back to the body, using methods involving acupuncture (a needle-based treatment), *moxibustion* (a treatment involved burning of the skin), and a wide variety of medicines. The Chinese also eventually led the way to the development of a process to prevent illness through vaccinations. As this chapter will show, the geographic isolation of China and the inability of ancient people to travel there regularly only slowed the release of news of the medical progress that was being made; it did not slow progress.

EARLY CHINESE CIVILIZATION AND THE ROOTS OF MEDICINE

Isolated from the rest of the world, the ancient Chinese proved to be ingenious people who utilized their country's vast natural resources

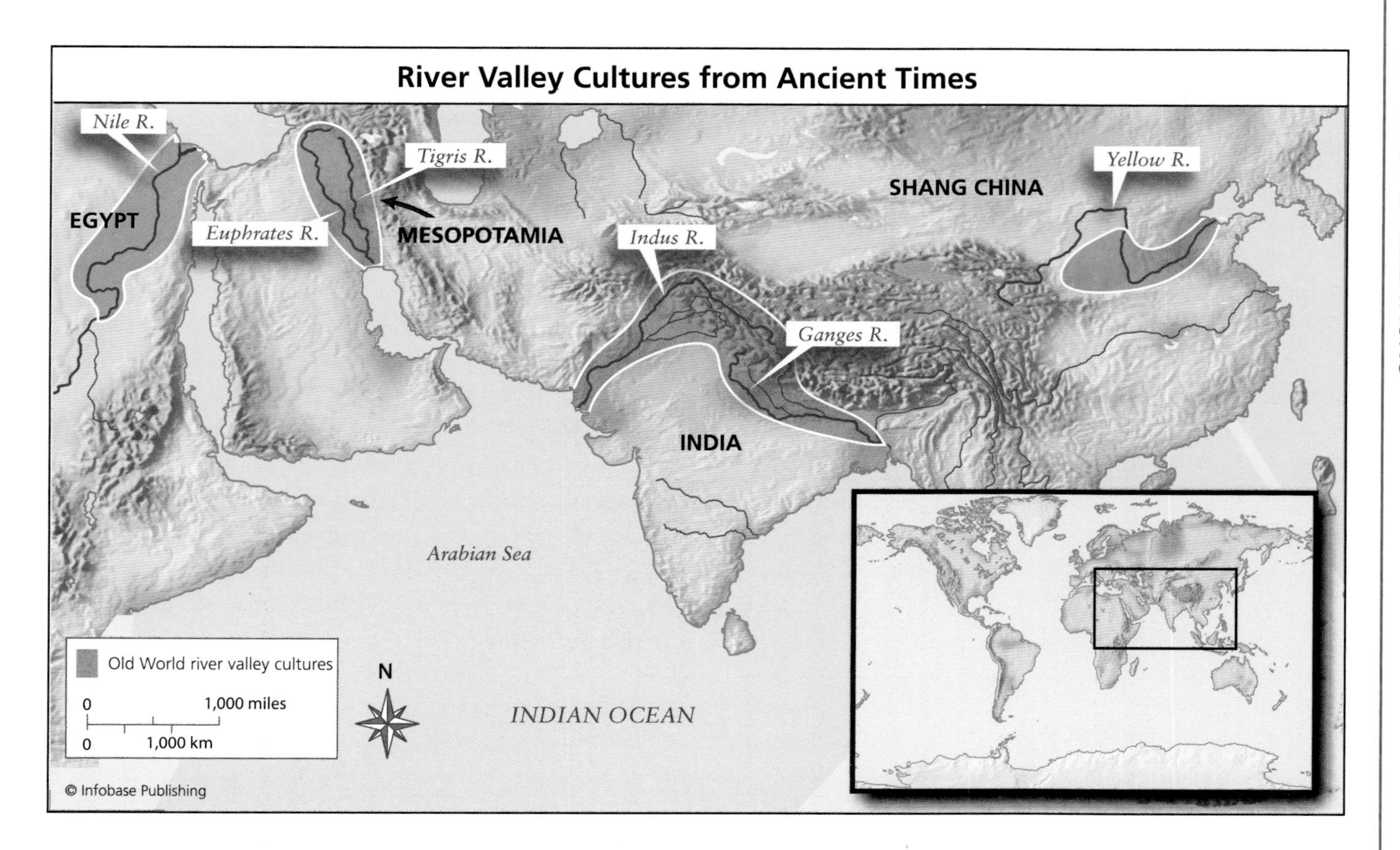
River Valley Cultures from Ancient Times
Nile R.
EGYPT
Euphrates R.
Tigris R.
MESOPOTAMIA
Indus R.
Ganges R.
INDIA
SHANG CHINA
Yellow R.
Arabian Sea
INDIAN OCEAN
N
Old World river valley cultures
0 1,000 miles
0 1,000 km
© Infobase Publishing

to address problems and to create solutions in many fields. In some cases, the Chinese developed inventions that the Western world did not discover for themselves until hundreds of years later.

Neolithic settlements along rivers in China date back to about 6000 B.C.E. By 1765 B.C.E., the Shang dynasty dominated the area, and they made advances that equaled those of Mesopotamia. Chinese knowledge of agriculture grew, and they developed the calendar to record planting times. Their invention of the chariot led to military campaigns that extended their reign over much of north-central China. The earliest Chinese writing known to exist dates to this time (1200 B.C.E.). Notations on bones (referred to as oracle bones) were found near present-day Anyang in Henan Province. Once deciphered, the bones contained information about hunting, warfare, weather, and ceremonies as well as recording information about illnesses and medical treatments.

During the Zhou (Chou) dynasty, Chinese society developed into a feudal system with landowning nobility whose local power superseded the dynasty rulers. A little more than halfway into this period (ca. 500 B.C.E.), the Chinese began to work with metal, learning to forge and cast iron. This development occurred in China a full 19 centuries before the rest of the world, and it made possible many improvements, from creating more effective farming implements to building stronger weapons.

Warfare and chaos dominated the Warring States Period (403–221 B.C.E.), and in 221 B.C.E. the Qin (Ch'in) dynasty came into power and put in place a strong central government. Its emperor also sought total control, ordering the destruction of all surviving manuscripts except for a few texts dealing with medicine, agriculture, and forestry. The dynasty did not last long, ending in 206 B.C.E., but from that date forward, China as a civilization took on a collective identity, and it was this dynasty for which China was named.

(Opposite) This map shows the development of ancient civilizations in major river valleys.

Although the unified force of a single ruling entity was not to return for many years, the Chinese people continued to make astounding discoveries that far outpaced their counterparts in the West. As early as 105 C.E., the Chinese created paper for writing, and their travel-related inventions such as the ship rudder in the first century C.E. preceded by 1,200 years anything similar in the West. The Chinese also developed the first magnetic compass and a system of cartography based on a grid system.

Religious influences were to come and go. Kong Qiu (K'ung Ch'iu, called Kongfuzi [K'ung-fu-tzu]; the name was Latinized to *Confucius* 2,000 years later) lived from 551 to 479 B.C.E. and was an important philosopher who was to have long-lasting influence on the thinking of the Chinese people. He pressed for justice for the people, and he looked for practical solutions, never supernatural ones. Although the teachings of Confucius went in and out of favor (both Buddhism and the simplicity of Taoism were popular at various times), Confucius was to affect medical thinking because the Chinese followed his desire for practical ideas for health and his advocacy for "medicine as a benevolent art."

The Roots of Chinese Medicine

The roots of Chinese medicine have been credited to three emperors: Fu Xi (Fu Hsi, ca. 2900 B.C.E.), who was said to have developed the concept of *yang* and *yin;* Shen Nong (Shen-Nung, ca. 2700 B.C.E.), who was thought to have invented acupuncture and discovered many of the medicines by personally testing them, sometimes as many as 70 in a day (his death was said to have been caused by a fatal sampling); and Huangdi (Huang-ti, ca. 2600 B.C.E.), who is supposed to be the author of *Huangdi Neijing* (*Huang-ti Neiching,* or *The Inner Canon of the Yellow Emperor*), the book that was to become the classic book on internal diseases.

But like other aspects of ancient history, more up-to-date information imparts a fresh view on old beliefs. An archaeological dig in 1973 at the Mawangdui (Ma-wang-tui) site (located in Changsha, China) contained human remains as well as texts that can be dated accurately to 168 B.C.E. The medical texts, which revealed practical experience with medicine, were notable for the information that was missing. There was no mention of acupuncture or moxibustion, leading modern-day scholars such as Imre Galambos, an expert in Chinese manuscripts at the British Library, to believe that the stories about the contributions of the three emperors are more myth than fact. The origin of some of these medical practices did not occur as early as was once thought.

From ancient times, the philosophy of Chinese medicine rests on classification and corresponding interrelationships among various parts of the body, as well as interrelationships to parts of the universe. Two methodologies of classification are used: yang and yin as well as a system that categorizes according to five elements, similar to the "four humor" theory of the Greeks. (See chapter 5.) Because of these interrelationships, the Chinese rule out the separation of mind and body.

The concept of yin and yang dates to at least the Shang dynasty (ca. 1750–1040 B.C.E.) and was rooted in the theory that everything within the body (and within the universe) must be in balance. Yang is thought to be associated with the heavens and related to exterior elements that are considered to be light and male. Yin is associated with the Earth and is the classification for those interior elements that are passive, dark, and female. Within the body, yang has to do with action and transformation. The large intestine, small intestine, gallbladder, stomach, and bladder are thought to be yang organs. Yin relates to circulation, nourishment, and growth and therefore is connected with the heart, liver, lung, kidney, and spleen. (The brain is not mentioned on either of these lists—most cultures did not yet understand its importance.)

The Chinese also classified things by fives, a method known as *wu xing* (*wu hsing*). Five basic elements (wood, fire, earth, metal, water) related to five directions, five seasons, five colors, five sounds, what were then the only known planets (five of them), and five organs of the human body. Each classification bore a relationship to something else, creating a dynamic system of interrelationships: For example, water puts out fire, fire melts metal, a metal ax cuts wood, a wooden plow turns up earth, and an earthen dam stops the flow of water. The organs were tied into this circle: The kidney (the organ of water) was thought to be in opposition to the heart, which was the organ of fire, and so on.

As part of this interrelationship, the Chinese believed that the body has a motivating energy that moves through a series of channels (also called *meridians*) beneath the skin. This is described by the Chinese as *chi* or (*qi*), which is defined as "life force." The Indian chakras are thought to be the same thing. (See chapter 3.) For thousands of years these cultures have studied, spoken about and utilized this energy form. If a person is feeling well, the energy flow is like a quickly flowing river. When this flow of energy along the circuitry (the meridians or energy channels) becomes blocked and throws off the balance of the *chi* or energy

THE PULSE

The pulse was so important in diagnosing illness that the early Chinese people were said to describe a visit to the doctor as going to "have their pulse checked." The Chinese believed that by checking the pulse, they could assess the circulation of the *chi* and correct any bodily imbalances.

The authority on the pulse was Wang Shuhe (Wang Shu-ho, ca. 180–270 C.E.), who wrote *Mai Jing* (*Mai Ching*), the Classics of Sphygmology, also known as the "Pulse Classic." This work was 10 scrolls long, and it identified 24 different types of pulse, from *fu* (floating) and *ti jie* (slow with irregular intervals to *cu* (running) as well as many others. Wang Shuhe (Wang Shu-ho) was the one who identified the wrist, where the radial artery can be felt, as an ideal site for taking the pulse. Shuhe's method involved checking the pulse in several places at different times and with varying degrees of pressure. The process sometimes took as long as three hours. As the science of pulse-taking advanced, Chinese physicians

within the body, then the imbalance can range from irritability to an actual illness.

DIAGNOSTICS IN ANCIENT CHINA

In ancient China, a diagnosis was not a simple process. Diseases were categorized by degree of sickness, and doctors were expected to distinguish among the various levels of illness. The early Chinese physicians studied for years to learn the appropriate diagnostic methods. Examination of a patient involved interrogation of the patient and the family to gain a case history of behavior patterns as well as current symptoms. They also employed all of the senses (visual, auditory, olfactory, and touch). Taking the

were expected to study 50 pulses and recognize more than 200 variations involving strength, volume, and regularity.

The importance of the pulse may actually predate Wang Shuhe. In what may be a mythologized tale (perhaps based on an actual circumstance), a well-respected physician, Bian Que (Pien Ch'iao), who lived during the Warring States Period (ca. 500 B.C.E.) was said to have been called to visit the crown prince of the Kingdom of Guo (Kuo), who was very ill and thought to be dying. By the time Bian Que arrived, the court physician had pronounced the prince dead, and preparations were being made for the funeral. Bian Que asked to see the prince's body, and during his examination he found a pulse. The prince had not died; he had gone into a deep coma. Bian Que applied special compresses, and within hours, the prince was awake again. Word spread that Bian Que could bring people back from the dead, but Bian Que humbly said that this was not the case; he had only treated him.

pulse of a patient was very important in making a diagnosis, and it was also complex. The early medical texts specified 51 different types of pulse that could be symptomatic of various health issues. (See the sidebar "The Pulse.") Inspecting the tongue was also vital to a diagnosis—medical texts noted 37 possible shades of tongue color.

Other recognized signs of illness included yellowing of the eyes, which indicated the ill health of the liver. (Today, this is known as jaundice, and modern doctors recognize this as a symptom of liver disease.) The surface of the ear was inspected to identify kidney disease; the lips were examined to see if there were problems with the spleen. The physician also noted the basic demeanor of the patient's face; he palpated the body (especially the abdomen) for tenderness; observed the sound of a patient's voice; compared the warmth and coolness of various parts of the body; noted any particular odors; and finally, asked the patient directly about the problem.

Because physical contact between physician and patient was kept to a minimum, patients were encouraged to indicate the location of their symptoms by pointing to an anatomically correct ivory figure that the physician supplied. The physician might take the pulse and look at the patient's eyes and tongue, but any other contact was kept to a minimum. When women had to consult a male physician, they were usually concealed behind a screen for the examination and used a servant or a husband as an intermediary to describe the problem and to indicate on the ivory figure the location of the pain or other discomfort.

THE HEALING ARTS: ACUPUNCTURE AND MOXIBUSTION

Even today, the Chinese are well known for two healing methods that date to the ancient era. Both acupuncture (a process of inserting fine metal needles into specific points along the body) and moxibustion (the burning of combustible plant material to create small blisters on the skin) were considered curative for both *somatic* (relating to the body) and *psychological* disorders.

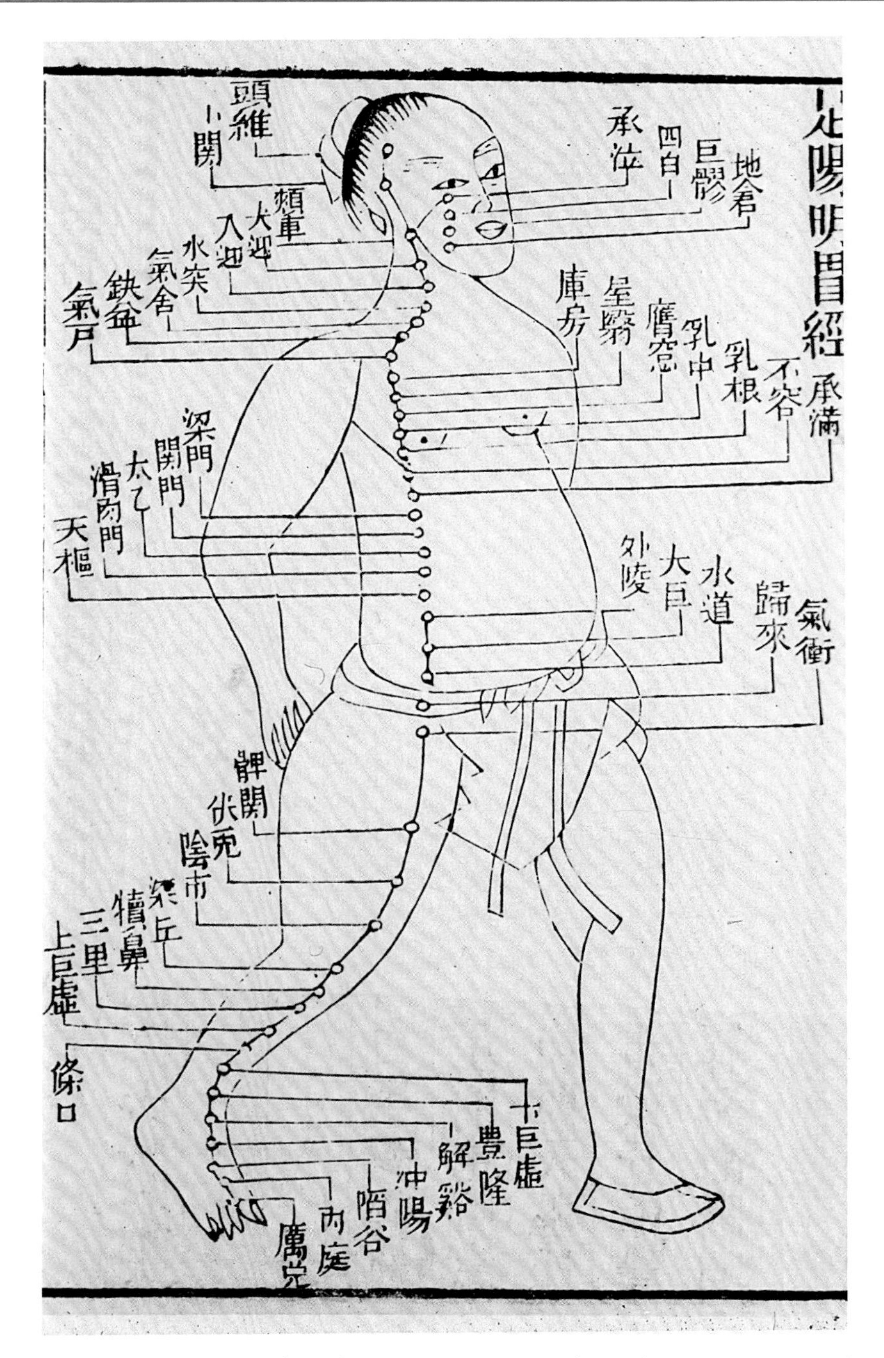

This is a depiction of ancient acupuncture points. *(National Library of Medicine)*

The exact date of the early use of acupuncture is a subject of debate, but by the second century C.E., Chinese medical texts noted 365 points along the body that were recommended as insertion points to help restore the balance of energy (*chi*) in the body and thus bring the person back to good health. The points, which are thought to correspond with various organs of the body, are located on 14 meridians running the length of the body; the belief was that disease came from an imbalance in energy flow.

The process is thought to have evolved from a time when the needles were used to drain pus or blood from an abscess. Even in early times, the science of acupuncture was complex and specific. Needles of various sizes were used for specific ailments, and they were inserted into the body in varying depths, depending on the nature of the energy imbalance. Once the needles were inserted, they were sometimes twirled or vibrated to restore the energy channels, and in this way, the acupuncturist could either extract or divert energy.

In moxibustion, a small quantity of combustible plant material was placed on the skin and set alight, causing a painful burn-blister that was intended to "counterirritate." The process was thought to provide heat to the body in such a way that it, too, would help restore balance to the body.

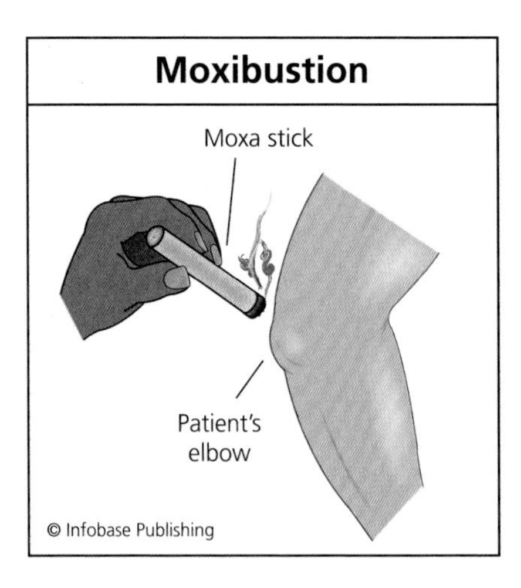

Moxibustion was a process that used heat or burning to stimulate circulation and induce smoother blood flow.

Moxibustion was generally recommended for chronic conditions such as tuberculosis, bronchitis, and general weakness as well as for a toothache, headache, or gout. In the seventh century, officials used moxibustion scars to protect against snakebites and foreign diseases before travel.

CHINESE PHYSICIANS

There were two ways to learn medicine in early China. The first was by studying under a healer, known as *a ruyi.* The other was through book learning and an apprenticeship that was generally passed down through a family of healers, known as *shiyi.* In addition to the scholar-physicians, there were also food physicians (focused on the healing powers of diet), physicians for simple diseases, ulcer physicians (surgeons), physicians for animals, and chief-of-physicians who supervised others. China had its share of laypeople who practiced medicine. Some were *charlatans,* but others studied acupuncture and became well respected for their art. Women served as midwives and wet nurses but were of low status.

Like other practitioners of ancient times, Chinese physicians preferred to treat those for whom the prognosis was good. Since physicians who failed to cure a patient were sometimes killed, most were reluctant to take on cases in which the prognosis was bleak. (The care of the very ill was left to inferior physicians.)

The following three particular physicians are known for their work during this early period:

Zhang Zhongjing

Zhang Zhongjing (Chang Chungching) (150–219 C.E.) wrote a book entitled *Shang Han Lun* (Treatise on Febrile Diseases) containing information on acupuncture, moxibustion, and herbal remedies. It is still used as a standard reference work for traditional Chinese medicine. Zhang Zhongjing also recorded a method of artificial respiration for the treatment of a suicide attempt by hanging.

Ge Hong (Ko Hung)

Ge Hong (283?–343 C.E.) also known as Ko Hung and Pao-p'u-tzu, was an alchemist from the Eastern Jin (Chin) dynasty who made great contributions to Chinese knowledge of infectious diseases. His work, *Zhou Hou Bei Ji Fang* (*Chou Hou Pei Chi Fang*) or *Emergency Formulas to Keep Up One's Sleeve,* includes valuable information about the transmission of disease, including tracing some

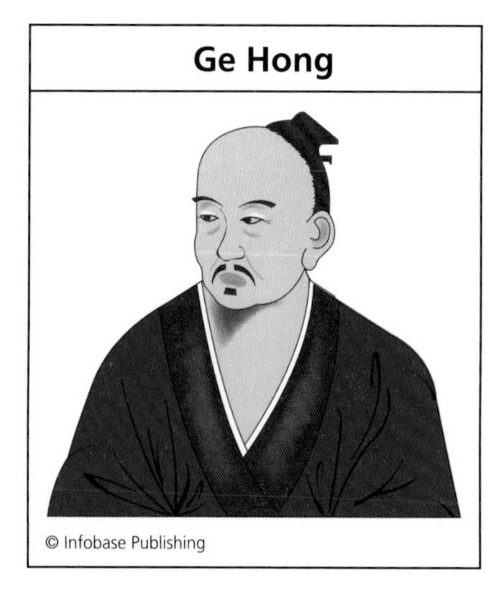

Ge Hong was the most famous alchemist in ancient China.

illnesses back to food and drink. He identified the sand flea as the cause of one particular illness (what the Chinese now know as *tsutsugamushi*). He also wrote of methods for getting rid of foreign bodies in the esophagus. Children must have swallowed fish hooks with some frequency, as Ge Hong (Ko Hung) developed a method for retrieval in which the physician would hold tight to a string of beads while pushing the beads into a child's throat. This must have forced the hook to pass down the esophagus. The beads were then retrieved by pulling on the string.

While *alchemy* is usually associated with trying to turn lead into gold, Chinese alchemists also experimented with drugs and herbal treatments in search of the great drugs of well-being and immortality. Ge Hong taught that minor elixirs could provide protection from ghosts, wild animals, and digestive disorders while superior elixirs could confer immortality.

"Like cures like" or "poison attacked by poison" was another of Ge Hong's concepts. For information on how his theory led to the creation of vaccinations, see the sidebar "The Creation of Vaccinations."

Hua Duo (Hua Tuo)

Surgery was not widely practiced in ancient China. *Dissections* were considered "mutilations of the body," so surgery was not taught, and it was practiced only under desperate circumstances. One Chinese physician, Hua Duo (Hua Tuo) (110–207 C.E.), who may have studied surgery in India before returning to China, made a notable contribution; he invented a method of rendering a patient temporarily unconscious. Hua Duo found that when an herb called *Ma Fu powder* was taken with wine, it worked as anesthesia. The West waited for more than 1,600 years before they had a similar way to remove a patient's pain during surgery.

Hua Duo's advanced knowledge and abilities brought about an untimely end, however. After Hua Duo successfully removed an arrow from the arm of General Guan Yun (Kuan Yun), he was summoned to another court for suggestions to help reigning prince Cao Cao (Ts'ao Ts'ao) overcome migraine headaches (ca. 265 C.E.). Hua Duo recommended trephination, but Cao Cao reacted badly to Hua Duo's suggestion of skull surgery. Cao Cao decided this remedy was actually an assassination plot, and he ordered Hua Duo to be arrested and executed. Cao Cao also required that all texts of Hua Duo be destroyed.

CHINESE PHARMACOLOGY (MEDICINE)

The Chinese *materia medica* (medicinal cures) has always been extensive, drawing from the vast resources made available by China's broad geography. The drugs themselves consisted of animal, vegetable, and mineral remedies and were carefully classified as to what they cured best. Some remedies were used to attack an invading agent, but more often it was to enhance the body's ability to restore harmony (yin and yang) so that a person could fight off illness himself.

Early Chinese pharmacology is thought to date back to legendary emperor Shen Nong (Shen-Nung) (2696 B.C.E.), the fellow who was said to have tasted hundreds of herbs to identify their medicinal properties. Since no written notations of that time have thus far been located, this may have occurred, or the remedies may have been passed down by word of mouth. The first true document of prescriptions, the Bencao Gangmu (Pen-ts'ao Kang-mu), probably dates to only about 200 B.C.E. It was said to contain about 240 vegetables and 125 other ingredients. At a much later date

(1597 C.E.), a government official known as Li Shizhen (Li Shih-Chen), spent 26 years cataloging herbal and drug knowledge and published a work often referred to as *Shen Nong Bencao* (*Shen Nung Pen Tsao*). The later work by Li Shizhen was 52 volumes and mentioned 1,892 drugs.

Some of the drugs would have been quite effective. Ma Hong (Ma Hung), a physician who lived during the second millennium B.C.E., recommended a drug derived from the Chinese joint fir *Ephedra sinica* for coughs and lung ailments, which is still used

THE CREATION OF VACCINATIONS

Though there are few details that explain any linear progression of this process, it is thought that the Chinese were the first to create a method of vaccinating against illness. There are not many accounts of how or why this process was discovered until one is reading about alchemist Ge Hong, who developed concept of "like cures like" or "poison attacked by poison." Ge Hong used some of the brain tissue of the rabid dog on patient's wounds as a treatment, and he noted that it was effective, protecting against the onset of hydrophobia (rabies) after an animal bite. This was the beginning of the concept of immunization.

Smallpox (the Chinese word for the disease translates to mean the "flower of heaven") presented a challenging puzzle for ancient people. The disease does not travel via any animal host so epidemiologically the virus responsible for smallpox requires a high population to spread. However, once it enters the population, the disease has a very high mortality rate. In time, court physicians in China realized that smallpox did not recur in those who survived it, and they began to experiment with ways to prevent the dis-

today (ephedrine). The Chinese used iron for anemia and Chaulmoogra oil for leprosy, a practice eventually used in the West. Ginseng was used to invigorate a patient and prolong life. Today, it is used for fatigue and also as a diuretic.

Other drugs were "match ups" and were unlikely to have worked as anything other than a *placebo.* For example, a red herb was used to treat the heart because it was red. Upper parts of plants were used to treat upper parts of the body. The Chinese also used a certain amount of what is referred to as dreckapothecary,

ease. They eventually found that scratching an unexposed person's skin with pus or scrapings from an infected person sometimes imparted immunity to the individual. Because the process involved variola virus, this method became known as variolation, though this type of vaccination process was eventually used for other illnesses as well. While this procedure sometimes prevented someone from becoming ill with the disease, it was not foolproof. There was a high risk of contracting the full-blown disease, and many died from the process.

By approximately the 10th century C.E., doctors created a dust by pulverizing the scabs from the smallpox lesions and prescribed that people should inhale it. By the 16th century, the Chinese were using variolation techniques similar to those introduced to Europe by Paracelsus, predating Edward Jenner (1749–1823), an English scientist who is routinely credited with creating vaccinations in the 1790s. (Medical documents from the 18th century indicate that Jenner's invention of vaccine was enlightened by the Chinese process of variolation.)

Pharmacy in ancient China *(National Library of Medicine)*

medicine from "dreck" such as remedies made of donkey skin, urine from various animals, human parts, and products such as human bones to treat syphilis. During the Ming dynasty, physician Wang Ji (Wang Chi) (1463–1539) claimed that he could cure syphilis with a preparation made from the bones of a dead infant.

Tea drinking became a popular Chinese practice, and in time, it assumed the reputation of promoting good health. Tea contains small amounts of nutrients but is rich in physiologically active alkaloids, including caffeine, theobromine, and theophylline. This process eventually led to a medicinal process known as decoction (the extraction of drugs by boiling certain herbs in water). Drinking a beverage that involved boiling—and thereby purifying—water probably was a main contributor to better health because it reduced the incidence of water-borne pathogen exposure.

CONCLUSION

China functioned in isolation for many years, so it was centuries before Chinese medical accomplishments were recognized. Only in recent years has Western medicine been giving more attention to the holistic medicine practiced by the Chinese; herbal treatments and acupuncture and a related form of touch therapy—*acupressure*—are being investigated by Western medical practitioners who are noting the limitations of the Western methods for disease-specific treatment. Their contribution to the understanding of the vaccination process was vital to the progress eventually made by Edward Jenner.

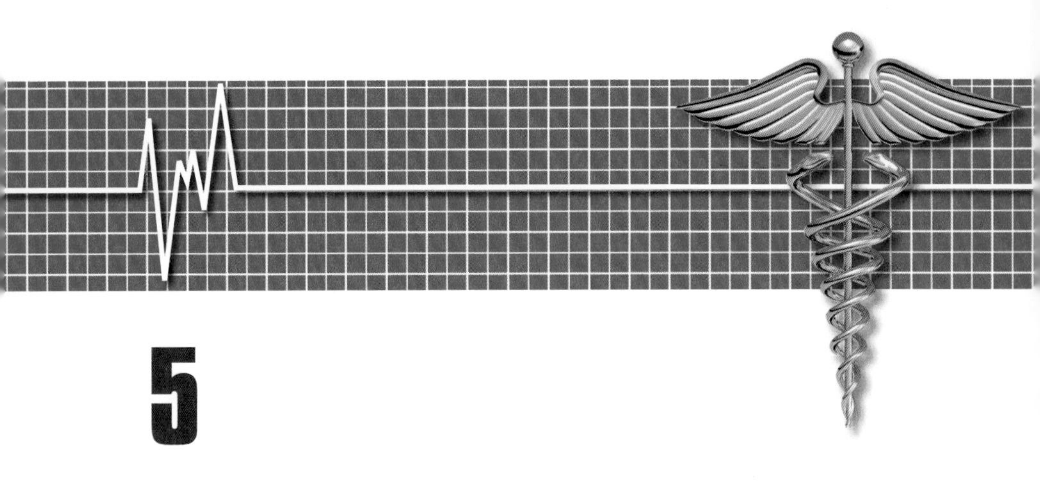

5

The Ancient Greeks and What They Learned

Just as today's historians are still arriving at new realizations about previous time periods, historians of the past did not begin their careers knowing that the Greeks led the ancient world away from religious healing and toward a more scientific form of medicine. Theodor Zwinger (1533–88 C.E.), a 16th-century Swiss physician and medical professor, was the first to examine the roots of early medicine and realize that the ancient Greeks saw that healing prayers had their limitations. Under the Greeks, the field of medicine became rooted in natural observation. While progress is never linear and the medical practices of the Greek culture slipped back and forth from science to prayer, progress was made overall. The ideas that came to the forefront during this period were to have far-reaching effects in the world of medicine.

Three overarching aspects of their culture had a direct effect on Greek thinking regarding health care. The first was the Greek emphasis on exercise and the well-toned body. As depicted in their art, the Greek ideal was a man with a muscular physique that he achieved through exercises and events involving physical prowess. As might be expected of a culture in which this was valued, athletic contests and festivals were held regularly throughout the land. As early as 776 B.C.E., the Greeks held events that were precursors to the modern Olympics.

Their response to the geography of the area was the second element of Greek culture that set the stage for change. Centralized government was impractical because of the divided geography of Greece. The residents of the loosely bound city-states of Greece were also skilled and comfortable seamen, so they willingly traveled and traded with the Egyptians and others throughout the Mediterranean. This exposed them to a wide variety of cultures and many new ideas.

As a people, the Greeks were also very open-minded about all that they saw, and this, too, changed their approach to medicine. The democratic ideals of Greece set a tone for questioning and exploring new ideas, and an openness to change was an important part of the Greek way of life. People of many professions were permitted to examine current practices and offer suggestions. As a result, the field of medicine expanded into new realms.

This chapter explores the root of early Greek medicine and the thinking behind the four humors that was the basis for medical belief from Greek times through the Middle Ages. Aristotle, who introduced the *scientific method,* and Alexander the Great, in whose namesake city was established a precedent-setting library for scholars, will be discussed. The chapter also sets the scene for the arrival of the Greek physician known as the Father of Medicine, Hippocrates.

EARLY GREEK MEDICINE

The epic poems *The Iliad* and *The Odyssey* are attributed to Homer, a blind Greek poet who lived sometime between the eighth and the sixth centuries B.C.E. These stories, which may have been part of an earlier oral tradition and written down at a later date, provide the earliest information about the practice of Greek medicine. While it is always difficult to separate fact from legend in stories, the plot lines of these epics include coping with epidemic diseases and treating battle wounds. These poetic descriptions provide insight into the state of medicine of the day and into the roles played by physicians, surgeons, priests, and gods.

Most of the treatment methods written about in this poetry were very practical. Though soldiers were unlikely to survive war injuries, it was not for lack of effort to relieve their suffering. The process often involved sucking the site of the wound to draw out poisons, thought of as the "evil influence," and then cleaning the area with warm water before applying medicines. The salves or drinking potions that were used at the time were simple ones that were derived from plants since medicinal concoctions were not of particular interest to the ancient Greeks. Sometimes, patients were given sulfur, saffron oil, opium, or just warm water or wine.

After hands-on practical solutions were attempted, prayers and incantations were used to supplement this care. Ailing Greeks were sometimes sent to visit the Temple of Apollo at Delphi. When a patient consulted Apollo's famous oracle, the oracle would go into a trance to receive instructions for healing. For years, this was thought to be a mythical description of the healing process at Delphi because no evidence led them to think otherwise. Then, in 2001, U.S. scientists discovered geological faults beneath where the temple once lay. Ethylene gas—later used as an anesthetic—may have been released through these faults, so the oracle may actually have entered a trancelike state. Team leader John Hale, an archaeologist at the University of Louisville, told reporter Bob Edwards in an interview on National Public Radio on February 15, 2002, that after taking samples of the springwater and the rock directly under the temple they found the answer: ". . . We found that there had been gases rising all through history along this fault, along with the springwater, and they were generated from the rock itself. It wasn't volcanic activity. This limestone was bituminous. It had petrol chemicals in it. And when the fault moved and the rock got hot from the friction of the moving blocks of rock, all those petrochemicals were vaporized, and the hydrocarbon gases, which are intoxicating—methane, ethane, ethylene—they all slid up the fault and emerged at the surface, right at the temple. So every detail of the geological setting supported the ancient tradition right down to the smallest statement about what was there."

As portrayed in these heroic poems, Apollo was both god-physician and god of prophecy. While he had the capability to

Head of Asclepius, copy after a statue from ca. 420–410 B.C.E. *(Bibi Saint-Pol)*

heal, he also could punish and was thought to sometimes send an epidemic to infect the people. *The Iliad* also introduces Asclepius, who came to be viewed as the god of healing by the fifth century B.C.E. In *The Iliad* Asclepius is described as the son of Apollo who was a tribal chief and warrior. He taught what was the Greek philosophy of healing: "First the word, then the herb, lastly the knife." Asclepius was thought to have inherited his godlike ability to heal from Apollo, and Asclepius's sons were known as Asclepiads, physicians from whom all other physicians were said to descend, and the names of two of his daughters provided the root for two health-related words in English today: Hygeia (hygiene and health) and Panacea (cure-all).

As time went on, the worship of Asclepius actually grew. While there were those who still believed in the more natural philosophy spawned by Hippocrates, the arrival of the plague increased the worship of Asclepius. (See later in the chapter for information on what happened during the plague.) There were not enough doctors to treat all who fell ill, and many could not afford treatment, so worship of Asclepius offered hope to those who had no other options.

Asclepius soon became associated with the snake, an important symbol of healing to Greeks as the shedding of snake skin depicts renewal of life. The snake was thought to have provided Asclepius with a precious healing herb. In art, a bearded Asclepius is depicted holding a staff that has a snake intertwined around it. A version of this, the *caduceus,* became the symbol of medicine and is still used today as the classic sign of the medical profession. The caduceus actually has two snakes intertwined on a winged staff.

A form of the caduceus has become a symbol of medicine. *(State University of New York, Oswego)*

GREEK UNDERSTANDING OF DISEASE

The Greeks believed that disease derived from an inner imbalance and not from outside pathogens. The key to good health was to maintain a balance of the bodily fluids, "the four humors" (see sidebar "The Four Humors"). Greek lifestyle valued exercise, good eating habits, and cleanliness, all of which were seen as ways to keep the body in balance.

The job of the physician was to observe sickness, attend the patient, and identify symptoms. To evaluate a sick person, a Greek physician went through a process involving inspection and palpation and the smelling of various fluids to ascertain what was wrong. Physicians also used an early form of auscultation (listening to the body) and succussion, a process that involved shaking a patient and listening to the fluids within the body. The Greeks also believed in "critical days" for illnesses: days 4, 7, 11, and 14 or 17 were the days that were considered turning points for different diseases.

As part of their belief in good health, dietary solutions were often prescribed. They viewed the first cook to be the first physician, and *dietetica* was a healing art that was linked to athletic training and to a well-regulated life. While other cultures used violent means of elimination, the Greeks favored more gentle bodily adjustments though they still used vomiting, purging, and *bloodletting* (the term *phlebotomy* meaning a surgical incision into a vein comes from Greek root words for "vein" and "cutting"). If mild methods did not cause the desired results, the Greeks then considered medicinal drugs. Surgery was used only as a last resort.

When challenged by endemic diseases like malaria that insidiously attacked the Mediterranean region, the early Greeks began to make connections between certain environmental elements and resulting illnesses. Marshes and malarial fevers were soon connected, and rats and mice were also known to bring about illnesses. While the true cause of these illnesses was not understood until the 19th century, this early understanding helped with lifestyle adjustments.

Most Greek physicians learned their craft through apprenticeship, though there were some written materials of the time providing clinical descriptions of diabetes, tetanus, diphtheria, and leprosy. Medicine was a male monopoly, though those who could enter the profession included a broad representation of social classes. Tradition permitted physicians to run fee-based services, but this then increased pressure on good results. Physicians only wanted to accept patients who would recover so that they would acquire more patients because of a good success rate. The physicians were known to remind people: "There is no skill where there is no reward." Nurses were used to help care for the sick, and midwives were consulted on issues regarding pregnancy and birth.

The Greeks did not have hospitals, so patients were either treated at their own homes, at the home of a physician, or at a shrine of Asclepius. Care for people of different social classes varied greatly. The poor were likely treated by servants of the doctor in much the same way as these underlings took responsibility for seeing to the needs of sick animals.

The Care of Women

The birth process presented major risks, and women commonly died in labor and of complications following the birth. The Greeks understood that the positioning of the baby was key to a successful birth, and much was written about helping to maneuver the infant into a helpful position. Cesareans are described in medical literature from this time, but the use of this term referred only to rescuing a baby whose mother had already died.

Women's issues were usually tended to by midwives, but sometimes women were so reluctant to discuss their condition that the problems became incurable. An apocryphal story, but one that concerned a very real issue, centered on Agnodice, an Athenian woman who was concerned that women were not receiving medical care because they were too embarrassed to discuss their problems with a man. She disguised herself as a man to study under Herophilus (see chapter 6), and with this training, she treated women patients. Although she was eventually discovered and prosecuted, her patients supported her. They argued that without Agnodice, they had no access to medical care. The prosecutors dropped the suit, and according to Hyginus, a Latin author of the first century C.E., the Athenians changed the law to allow free women to study medicine. Modern scholars view this story with some skepticism. Male doctors were increasingly becoming involved in gynecological care, so the story runs counter to what may have been happening at the time, though it is illustrative of an issue—the reluctance to discuss intimate matters—with anyone.

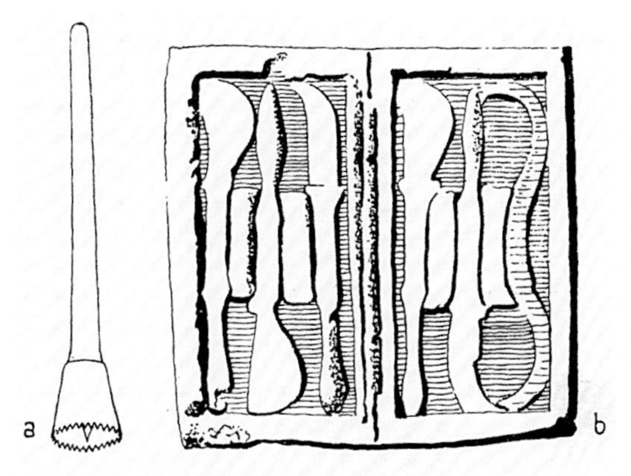

a
b
FIG. 15. Types of instruments used by Greek surgeons
(a) Simple trephine with centre pin. (b) Case of scalpels.
(a) Sixteenth-century instrument of ancient type. (b) Relief in the Asclepieion, Athens.

Ancient Greek surgical tools *(modeled after Singer, Charles, and Underwood, E. Ashworth,* A Short History of Medicine*)*

ARISTOTLE'S INFLUENCE

A long line of philosophers and teachers contributed to the Greek approach to medicine. Socrates (470–399 B.C.E.) was the first to develop a method involving an orderly series of questions for approaching any type of problem. This Socratic Method led the way to the eventual development of a scientific method of inquiry. Plato (427–347 B.C.E.), his pupil, learned from Socrates and then took responsibility for documenting some of Socrates' teachings. In turn, Plato's own pupil, Aristotle (384–322 B.C.E.) continued the writing tradition taught by Plato and went on to be a great physician, philosopher, and biologist in his own right. Along with Socrates and Plato, Aristotle was one of the most influential ancient Greek philosophers. These men transformed pre-Socratic Greek philosophy into the foundation of Western philosophy and ethics that have had lasting influence on the world today.

Aristotle studied widely and had a major influence on most fields he studied. He was a proponent of scientific, systematic observation and experimentation, which further advanced the creation of what was to become known as the scientific method. His work has influenced medicine and science for 2,000 years.

In addition to his philosophical writings, Aristotle pursued other fields of study and was one of the first Greeks who believed in the importance of understanding the anatomy. By the end of the fourth century B.C.E., he and his followers embarked on a major effort to use animal dissections to arrive at a better understanding of zoological and biological matters. His studies permitted him to observe the heart, but it was left to others to discern the four chambers and distinguish between arteries and veins. Though it took almost 2,000 years, Aristotle's work laid the groundwork for the English physician William Harvey (1578–1657), who developed the first complete and correct theory of the circulation of the blood.

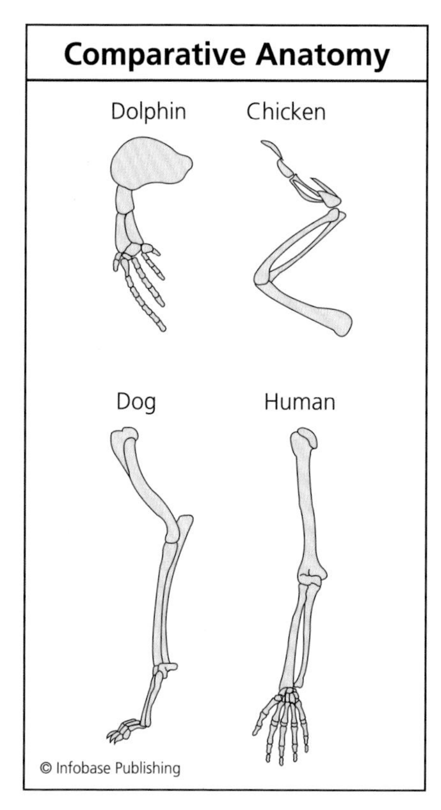

Aristotle and his students began to study different animal dissections to understand human anatomy better.

Aristotle also studied meteorology, metaphysics, *cosmology,* ethics, politics, and poetics. He eventually became tutor to Alexander the Great.

THE ALEXANDRIAN SCHOOL AND ITS EFFECT ON MEDICINE

Alexander the Great (356–323 B.C.E.) was one of the most successful military commanders in history. He inherited from his father, Philip II of Macedon, a unified group of Greek city-states. (See the related sidebar "Alexander the Great" concerning the medical theories concerning Alexander's own early death.) When Philip died, these city-states promptly rebelled against having a new leader, and Alexander had to reestablish family domination over the territory. He accomplished this and more, overtaking the Persian Empire, including all of the Mediterranean, Anatolia, Syria, Phoenicia, Judea, Gaza, Egypt, Bactria, and Mesopotamia. As he enlarged his territory, he ushered in what is known as the Hellenistic era of Greek civilization, which became an important time for the development of science and medicine.

Ultimately, this broad expansion of territory led to the collapse of his empire as the local rulers became power-hungry and carved out their own kingdoms. Both Pergamum (birthplace of Galen, a physician who was to exert great power in the field of medicine) and the Ptolemies in Egypt were two kingdoms that resulted. King Ptolemy (323–282 B.C.E.) established court at Alexandria at the mouth of the Nile, and he created the Alexandrian Library and the Museum (sanctuary of muses) that brought Greek learning to a new level. Because Alexander's conquests had opened new territories, the scientists at the library benefited from being able to study previously unknown animals, plants, and minerals. Eventually, the Alexandrian Library had 700,000 manuscripts, an observatory, zoological gardens, lecture halls, and rooms for research.

The ongoing studies at the Alexandrian Library gave rise to changes in the way doctors practiced medicine. Among the developments was an empirical method of practicing medicine, whereby physicians worked from observation and experience. Eventually, the three fundamental steps that were part of this process were codified: anamnesis (the taking of a case history); autopsy (which referred to visiting the ill and inspecting the patient), and diagnosis (an assessment of the illness). While this process was more practical than was worshiping a god or casting spells, it still lacked true scientific understanding of disease. As a result, the patients' outcomes often were not successful.

The Alexandrian physicians benefited from Egyptian influence. The Egyptian accepted cutting into the body as part of the embalming process, and this opened Greek minds to the possibility

THE FOUR HUMORS

Ancient people were always looking for ways to explain the world, and one of the early and lasting theories had to do with the balance of four elements: air, fire, water, and earth. This idea was introduced by Empedocles of Agrigetum (490–430 B.C.E.) as a way to explain the universe. Soon it became a major theme of medicine.

The concept of "nature in balance" became an important part of Greek culture. Hot and dry, and wet and cold were eventually added into this understanding of nature as were the four seasons, four ages, and eventually, four temperaments. Soon, the Greeks used the four humors to explain why some people became ill when others did not. Hippocrates was the first to popularize the theory of keeping the four bodily humors (blood, yellow bile, black bile, and phlegm) in balance. Later on, physicians determined that the humors related to heart, brain, liver, and spleen. Cures of illnesses were based on evaluating an imbalance and then working with an opposite property to bring the body back into balance. For example, black bile was considered to be dry and cold, so if this was what was judged to be in excess, then it was treated with hot and wet remedies. (Yellow bile caused warm diseases; phlegm caused cold ones.)

Many theories surrounded the four elements. Physicians differed in their thinking as to how the imbalances occurred. Some thought the body created these substances, and they

of dissection. For the first time, the ban on dissection was lifted. Unlike the Egyptians, who did not study the body as they cut and embalmed, scientists in Alexandria viewed systematic dissection for the essential information provided. They also established that this knowledge was valuable even if it was not immediately appli-

waxed and waned within the body. Others thought it was the result of something that an individual consumed. Warm foods were thought to produce yellow bile, for example.

A Greek philosopher, Theophrastus (372–287 B.C.E.), a pupil of Aristotle's, added to the thinking about these elements by suggesting certain personality characteristics that related to these elements. If a person had too much blood, they were sanguine; if they had too much phlegm, they were phlegmatic; if they had too much yellow bile, they were choleric; if they had too much black bile, they were melancholy. All of this was eventually incorporated in the Hippocratic writings (see chapter 6 for more information) that were further developed by Aristotle and Galen of Pergamum (see chapter 7), and the theory of the four humors became the rationale behind medicine of the Middle Ages.

The theory of the four humors influenced Western medicine for more than 1,500 years until the 19th century. Galen, a very influential physician, was among its strongest proponents, which increased the credibility of the idea. No one else emerged with any good alternatives, and because the theory made a certain degree of sense—a person suffering a cold produced phlegm; a person with an upset stomach created bile—the theory held sway for 1,500 years. It was not replaced until 1858, when the German physician and pathologist Rudolf Virchow (1821–1902) developed his theories of cellular pathology.

cable. (For more information on the Greek healers who led the way in undertaking dissection, see chapter 6.) Though the practice of dissection was sanctioned, it was still controversial. Some leaders of the day felt that anatomical research distracted physicians from care of the patients.

The Alexandrian Library lasted for several centuries, and many were trained in the fields of science and medicine. After Julius Caesar conquered Egypt, Alexandria was no longer the seat of intellectual learning, but the museum and the library continued to operate on a less grand scale. Political turmoil was brewing, and when Caesar arrived in 48 C.E., riots erupted. Parts of the library may have been destroyed at that time. Later, Christian leaders encouraged the destruction of pagan temples, and the library and the museum may have been further harmed in this warlike fervor. The story is told that the last scholar still working at the museum was a female philosopher and mathematician named Hypatia (ca. 350–415 C.E.), who was waylaid by a Christian mob and beaten to death. Muslim conquest of the city between 642 and 646 C.E. resulted in the final destruction, bringing an end to the broad and remarkable collection of books and manuscripts that had lasted until this time.

THE SPREAD OF THE PLAGUE

During the early fifth century B.C.E., the Greek military ran several successful campaigns, taking over much of the land in the Persian empire (490 and 480 B.C.E.). In the aftermath, several Greek cities united with Athens to create a unified league for defending the seas, but during that time, animosity among the united cities began to brew. Eventually, Sparta launched an attack on Athens. This was the opening volley in the Peloponnesian Wars (431–404

(Opposite) The Greeks were excellent navigators, and as they sailed the Mediterranean, they carried with them their growing body of knowledge about medicine.

Ancient Greeks and the Spread of Medical Knowledge

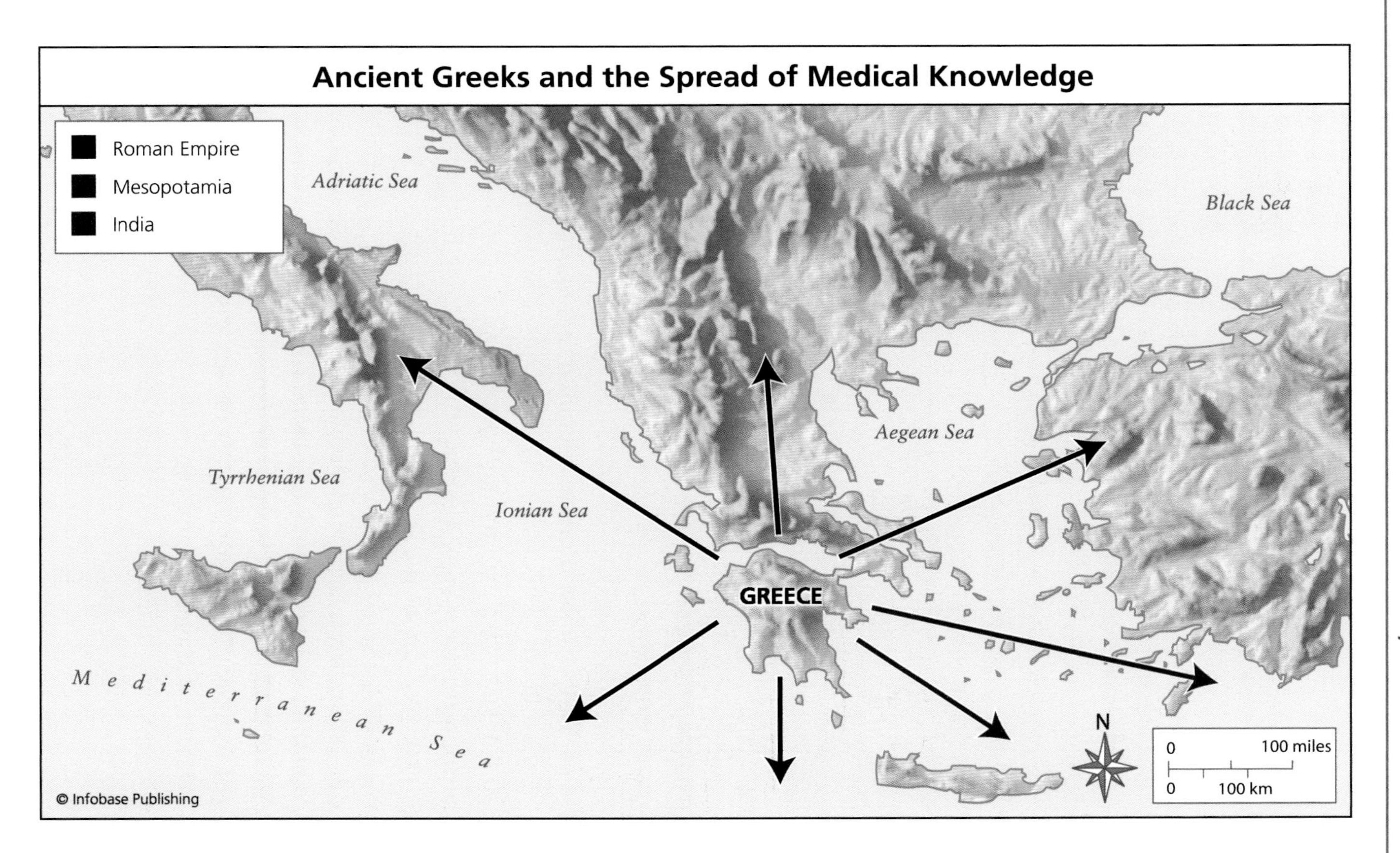

ALEXANDER THE GREAT (356–323 B.C.E.): *The Death of a Great Leader*

The death of Alexander III, the son of Philip II of Macedon, was illustrative of the state of medicine of the day. By the age of 33, Alexander the Great, as he was known, had conquered most of the world that was known to the Greeks at that time, and he was undefeated in battle.

The year of Alexander's death is certain, but the cause of the death of such a strong warrior in the prime of life is not. Originally, three theories were given: poisoning, overdrinking, and a relapse of malaria (he had contracted it originally in 336 B.C.E.). They know that he spent many nights out drinking despite a fever that he contracted.

More recently, another theory has been advanced: Alexander may have died from a cure rather than a disease. Some speculate that he took too much hellebore, a plant substance used as a purgative at that time. It was thought to be medicinal but was poisonous if taken in large quantities. At this time, there two kinds of hellebore: black hellebore and white hellebore. The second is highly toxic but is believed to be the one that Hippocrates used in small doses as a purgative. Black hellebore is also toxic if too much is taken, but it was still used to treat paralysis, gout, and other diseases.

While there is no conclusive evidence for one cause or another, it is relevant to note that the powerful and well-respected patients were the experimental "guinea pigs" of the day. Only very basic care was given to the poor, so wealthy individuals and the ruling class were subjected to "cures." In many cases, these cures were ineffective and sometimes deadly.

B.C.E.). This Greek military conflict rose between the Athenian empire and the Peloponnesian League, which included Sparta and a group of cities. This was a significant turning point in Western civilization as well as the backdrop for one of the fiercest plagues in history.

In the second year of the Peloponnesian War (430 B.C.E.), the Greek historian Thucydides (date of death: ca. 401 B.C.E.) described the coming of an epidemic. The spread of disease began in Ethiopia, passed through Egypt and Libya, and then came to the Greek world. The disease itself was exceedingly virulent. Formerly healthy people suffered headaches, sneezing, hoarseness, pain in the chest, coughing, and vomiting with violent spasms. Sufferers generally ran low fevers, but, eventually, they showed signs of blistering and ulcers. Some lost fingers, toes, and eyes; some suffered from severe mental confusion. Death usually came within seven to nine days.

Because of the ongoing war and the attacks from the Spartan military, Pericles (495–429 B.C.E.), an Athenian political leader of the day, called the people from the outlying agricultural areas to come within the city walls for protection. As the farmers crowded into the city, food production dropped, and the additional density of the community created a perfect environment for the spread of disease. During the summer of 430 B.C.E., the disease spread throughout the area, with people dying in large numbers. The plague returned twice more, in 429 and 427–426 B.C.E. (In 429, it took the life of the Athenian leader himself, Pericles.)

Thucydides' detailed descriptions have been helpful to historians, and he describes social disintegration as a difficult part of the aftermath. No one wanted to treat the suffering for fear of becoming ill themselves. Eventually, they found that the body built immunity to the disease, so those who recovered were sometimes willing to help the ill.

While there is much speculation as to what the actual disease was—smallpox, measles, typhus, and even syphilis have been suggested—scholars are quite clear that this plague was not the

Black Death (the term used for the bubonic plague that struck Europe in the 14th century). What is known is that it killed 25 percent of the Athenian soldiers and approximately 25 percent of the civilian population. As a result of the plague, the possibility of a Greek empire died.

The plague also demonstrated that there were too few physicians with too little knowledge to do much for this crisis. Hippocrates was said to have counseled to patients, *"cito, longe, tarde,"* which translates as "go fast, go far, return slowly," in response to the plague. The need for healing meant that the Greeks returned to prayer and beseeched the Greek god Asclepius to help their loved ones. According to *The Iliad,* it was during this crisis—during the age of Hippocrates—when Asclepius achieved godlike status.

There were no hospitals, and physicians were expensive and not readily available, so people made pilgrimage to the nearest temple. A belief developed that sleeping in the temple, called incubation, was healing. Sometimes, preliminary preparation for this visit included fasting, self-mutilation, or taking hallucinogenic potions. The sick person would sleep in the incubation chamber in the temple, and his or her dream would suggest a cure. If the solution was not immediately clear, the temple priest would interpret the dream and explain what the patient must do. The fortunate recovered. There was no charge for treatment at the temple, but patients were expected to leave offerings afterward.

By the third century B.C.E., the culture of Asclepius spread, and by 200 B.C.E., every large town in Greece had an Asclepian temple. Later, the "temple cure" became popular among the Romans as well, and people came for all types of help, from ways to increase fertility to potions for growing hair.

CONCLUSION

The next chapter will describe the contributions of the renowned Greek physician Hippocrates, who came to be regarded as the Father of Medicine because of his belief in the need for patient observation and for careful scrutiny of the symptoms of the disease. He also encouraged natural healing.

From 250 B.C.E. forward, Greece began to fall increasingly under the military influence of Rome. Rome, in turn, began to be influenced by the ideas of the Greeks. These ideas were then carried forward as the Roman Empire expanded. By 80 B.C.E., Greek physicians and their theories were common in Rome and throughout much of the region.

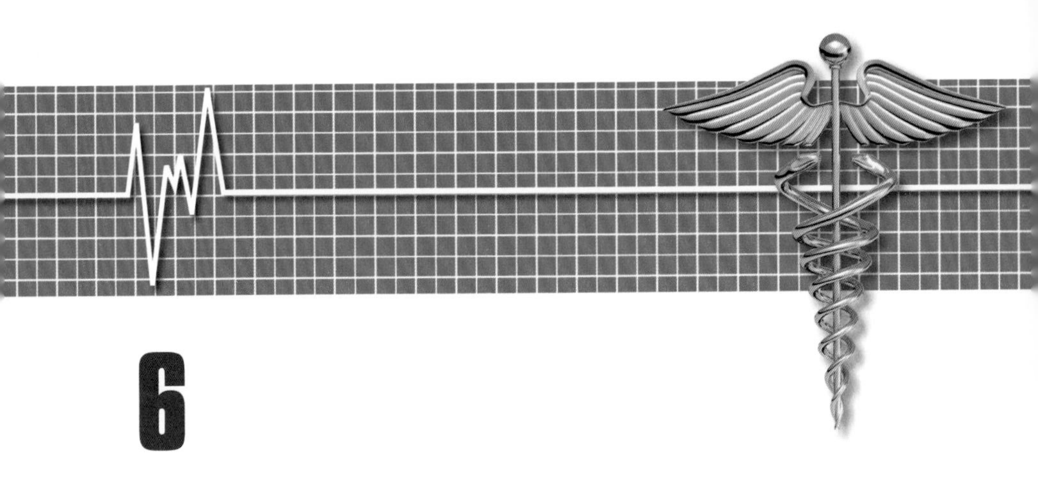

6

Hippocrates and the Importance of Medical Observation and Documentation

The dawn of a new understanding of medicine in Greece began in the fifth century B.C.E. with the teachings of Hippocrates of Cos (460–377 B.C.E.). Hippocrates introduced the concept that diseases come from natural causes, and he strongly rejected the Greek healers who recommended prayer as the prime methodology to make people better. He also founded the Hippocratic School of Medicine, which provided him with many followers who furthered his beliefs and practices.

Contemporaries regarded him as a brilliant man who contributed greatly to the field of medicine, and as time went on, he came to be thought of as the Father of Medicine. While some medicine is very disease-focused (the practice of Western medicine today), Hippocrates believed the well-being of the patient needed to be the physician's focal point; he also taught that healthful lifestyle choices were the first steps on the road to recovery.

Hippocrates taught that by undertaking careful observation of the patient and his symptoms, a physician would better understand the disease. He and his followers were the first practitioners to document carefully what they observed. This patient-oriented systematic method of observation and treatment eventually became the foundation of Western medicine.

While Hippocrates did a great deal to create a more scientific approach to medicine, the primitive state of science coupled with the fact that the Greeks disapproved of anatomical dissection meant that Hippocratic medicine involved a certain number of misunderstandings about causes and cures of disease.

For many centuries, it was thought that Hippocrates alone was responsible for the contributions that date back to this period of Greek medicine. As scholars have examined the material more critically, they have determined that the body of work originally attributed to Hippocrates is not the work of one individual. The *Hippocratic Corpus*, as it is known, involves various writing styles as well as some contradiction of belief. Today, the documents can be dated more accurately using the latest technology, and the works are now thought to have originated in a period of about 50 years (approximately 420–370 B.C.E.). The treatises are now thought to have been written by a number of physician authors, and exactly what Hippocrates wrote—and what he thought—is a subject of ongoing debate.

Hippocrates is widely known for the Hippocratic Oath, a statement of professional conduct. In time, Hippocrates came to represent the dignity of Greek medicine. Based on the latest information, the Hippocratic Oath is now no longer believed to date as far back as Hippocrates. His influence, however, set the tone for this code of conduct.

Very little is known about Hippocrates' life. The first biography about him is thought to have been written almost 500 years later by the Greek physician Soranus (ca. 98–138 C.E.). He is thought to have traveled widely practicing medicine, and he taught his theories at the medical school he founded in Cos. Because there are no contemporary descriptions of him, scholars are left to speculate. There is general agreement that he was an intelligent and gifted practitioner; various accounts describe him as kindly, while others attribute to him a sternness of presence. Certainly, the fact that he was patient-focused might lead to the thought that he meant well by his fellow man.

This chapter will examine the new direction in which Hippocrates led medicine and introduce the various medical schools of thought that followed.

A RATIONAL APPROACH TO MEDICINE

Hippocratic medicine featured the following particular elements, though they were only inconsistently practiced by others during this period of time:

- an emphasis on paying close attention to a patient and his or her symptoms;
- a belief that diseases came from natural causes (often imbalances) not as a result of a god's punishment; and
- an open-minded attitude toward the ideas of others.

This patient-centric view of medicine is still part of some aspects of medicine today.

The shortcomings of Hippocratic medicine stemmed primarily from the lack of knowledge about the human body. Since dissections were not undertaken until the third century B.C.E., Hippocrates and his followers did not fully comprehend how the body worked. They learned what they could by examining wounds and making educated guesses as to how the body worked, but there were big gaps in their understanding. A few physicians performed dissections on animals, but this led to errors in thought because both human and animal anatomy differ in some significant ways.

Under Hippocrates' tutelage, Greek medicine became very patient-centric. A physician's first task was to create a profile of the patient, outlining where they lived and worked, their way of life, and what they ate. Much of this information was elicited through the asking of questions; yet the physician also used all of his senses to observe the patient. A Greek physician prided himself on the ability to identify helpful signs to predict the course of an illness. In time, Greek physicians kept careful records about various diseases, and being able to refer to this body of work improved their ability to develop a prognosis. This was a very important part of Greek medicine. Like the Chinese practitioners, a physician could not afford to tarnish his reputation by having patients die. If the outlook for a person was bleak, many physicians would refer the patient elsewhere. Miracle cures were never promised, and the Greek intent was to prove to be a "friend to the sick" who would do no harm.

THE HIPPOCRATIC APPROACH TO TREATMENT

"Our natures are the physicians of our diseases," is attributed to Hippocrates and well sums up the approach to the medical treatment he advocated. The prime emphasis in any type of treatment had to do with reinstating the balance of the humors. Treatment of the whole patient was also a vital component in Hippocratic medicine. The physician who treated just the disease or only one body part would not be successful.

Hippocratic physicians recommended adequate rest and exercise, with the thought that these efforts would encourage the body to self-heal. Improvement of diet was also important, but the first measures taken to cleanse the system involved a period of near starvation. The patient could consume only honey and water, or sometimes vinegar, honey, and water. If phlegm was a major symptom of a particular illness, citrus was sometimes recommended to help rid the body of an imbalance of phlegm.

Various types of bleeding, including nosebleeds and menstruation, were considered helpful in restoring some imbalances, and bloodletting was also practiced to "rid the body of excess blood." If these methods failed, the patient was sometimes given some type of medicinal drug to cause vomiting or purging of the system.

Hippocrates preferred mild remedies, but other physicians prescribed an array of drugs, often concocted from local plants but sometimes using imported ingredients from Egypt and India. One well-known diuretic was made of a certain type of beetle with its head, feet, and wings removed. These medicines were applied as poultices and ointments or given as pills or suppositories.

Hippocrates, great Greek physician of the fifth century B.C.E., is pictured palpating a young patient. Kindliness and concern, embodied in his aphorism, "Where there is love for mankind, there is the love for the art of healing," are reflected in Hippocrates' face. *(Department of Library Sciences, Christian Medical College—Vellore, History of Medicine Picture Collection)*

In general, Hippocrates believed in doing what he could to aid in healing, and then it was important to let nature take its course. This was particularly helpful with the healing of broken bones. Physicians could put a patient in traction or reset or splint a bone. Time would do the rest. One of the treatises notes that a broken leg should be expected to heal in about 40 days.

Surgery (the word is from *chirurgia*, a Latin word that comes from the Greek *cheiros*, meaning hand, and *ergon*, meaning work) was considered to be beneath the work of a physician, so it was either left to others or practiced as a last resort. Hippocratic medicine specifically forbade physicians from "cutting for stone," a practice for relieving kidney stones that was used by other ancient cultures. Though the followers of Hippocrates did not perform surgery, it interested them, and they studied and wrote about it.

IN THE WORDS OF HIPPOCRATES

Hippocrates is credited with many wise comments about the practice of medicine, and many of sayings have been collected in a book called *Aphorisms.* Because they were originally written in Ionic Greek, the exact quote is dependent on the translator. The following are some of the aphorisms that are attributed to Hippocrates:

- *"Primum non nocere."* First, do no harm.
- "Life is short, [the] art long, opportunity fleeting, experiment treacherous, decision difficult."
- The art of medicine has "three factors, the disease, the patient, the physician."
- "What drugs will not cure, the knife will; what the knife will not cure, the *cautery* will; what the cautery will not cure must be considered incurable."
- "When the disease is at its height, it will then be necessary to use the most slender diet."
- ". . . When during the same day there is at one time heat and at another time cold, the diseases of autumn may be expected."
- "And in whatever part of the body there is a sweat, it shows that the disease is seated there."
- "Persons whose noses are naturally watery, and their seed watery, have rather a deranged state of health; but those in the opposite state, are more favorable."

Though this information may or may not have come from Hippocrates, the advice was notable enough to live on as recommended practices from that time.

HIPPOCRATIC CORPUS

The *Hippocratic Corpus* is the term used to describe 50–70 books that for centuries were all attributed to Hippocrates. The works, including textbooks, lectures, research notes, and philosophical essays, were written on various aspects of medicine for both professionals and laypeople. In the Alexandrian Library, they were collected and identified as the *Corpus Hippocratum.* On closer study, historians noted that the texts consisted of multiple writing styles. Some works are actual texts, some are speculative essays, and others are written as case notes. The current belief among experts is that the writings range in date from 420 to 370 B.C.E. and that they represent the work of multiple authors.

Though there are many contradictory statements within the various works, the Hippocratic books are united by the focus on Hippocratic medicine and the belief that people are governed by the same physical laws as the cosmos. The authors, who remain nameless followers of Hippocrates, express strong support for explaining health and disease via empirical and rational reason-

Hippocrates refusing the presents of Artaxerxes *(Heritage History)*

ing; they vociferously reject the idea that disease was caused by the vengeance of the gods. Most of the texts emphasize the value of observation in understanding the disease process and developing a prognosis. The authors stress the importance of a patient's overall well-being and do not recommend focusing on just one part of the body. These books, too, state that treatment should be offered only to those for whom there is hope.

The wide variety of issues addressed by the works of the *Corpus* is illustrative of the thinking of the day. The following table provides some short descriptions of a few of the works:

SOME OF THE WORKS BY HIPPOCRATES AND FOLLOWERS	
On Ancient Medicine	A very important text. The healing component of nature is the primary thesis. The job of the physician is to work with nature to create a harmonious balance. Dietetic observations and solutions were frequently a part of the process. The book also cites recognizable symptoms of certain diseases and notes that climate can affect health and disease.
Airs, Waters, Places	The first classic of medical geography. The text reveals that the physicians of the day had come to the understanding that diseases such as malaria were seasonal. The authors wrote of the effect of weather and attributed the season as causative, though they did not yet seem to understand that mosquitoes were a part of the disease-spreading process. Physicians of the day frequently traveled from town to town offering services where needed. *Airs, Waters, Places* notes that when a physician arrived in a new locale, he needed to strive to understand the area and its climate. Then he would be better prepared to understand the local disease. The second part of the work focuses on health and disease issues in parts of Europe and Asia.

(table continues)

SOME OF THE WORKS BY HIPPOCRATES AND FOLLOWERS

(table continued)

On the Nature of Man	This is an anatomical and physiological treatise that presents the four humors and embraces them as part of the medical process and establishes their correspondence to matter—air, water, earth, and fire. It presents blood as necessary for the life of humans and animals and compares it to the sap of plants; phlegm, bile, and black bile are all related to various illnesses.
On the Sacred Disease	Focuses on the symptoms and treatments of a particularly fearful disease of that time—what is now known to have been epilepsy. The Hippocratic author resoundingly writes that it is not a "sacred" or a divine desease, and the text makes a strong plea for a natural explanation of disease as opposed to supernatural. At the time, it was thought that epilepsy was caused by an imbalance—a blockage of phlegm in the airways. The convulsions that are part of the illness were the way that the body attempted to open the blocked pathway. This book also notes the importance of the brain, which was a new recognition of the time.
On Prognosis	Provides intimate knowledge of symptoms possessed by the Hippocratic physician and gives highly sophisticated descriptive detail.
Epidemic Diseases	Written about diseases on the island of Thasos, and the descriptions are notable for their clarity. Modern physicians can identify many of the disease described.

But after the *Corpus* there was silence for a time. From the *Corpus Hippocratum* to the work of Roman Celsus (about 30 C.E.), medicine made progress but lost some of the clear and simple approach to treatment.

THE HIPPOCRATIC OATH

Hippocratic Oath in the Shape of a Cross

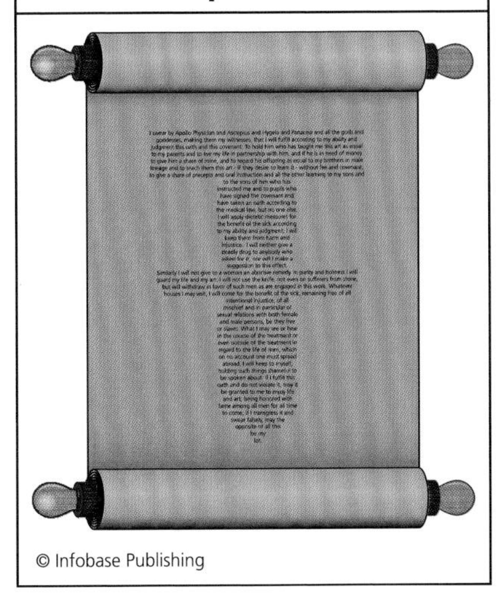

© Infobase Publishing

The Hippocratic Oath is one of the oldest, binding documents in history, and as a result, it has been used in many different ways, including being depicted in artwork.

Though the Hippocratic Oath is one of the most famous pillars of medicine and ethics, there is little information about its origin. Today, there is uncertainty about the date on which it was composed, its intended purpose, and exactly who put it together.

The oath consists of two parts—the first section talks of the duties of a pupil to his teacher and his obligation in transmitting medical knowledge to others, and the second part has a short summary of medical ethics as a self-regulated discipline. One of the reasons that there is doubt as to it originating with Hippocrates is that it is not fully consistent with Hippocrates' own theories. Both suicide and *abortion* are specifically prohibited by the wording of the oath, but both were practices of Hippocrates' day. Abortion (the prescribing of a "destructive pessary") and infanticide were both practiced during ancient times as a means of birth control. During Pythagorean times, it became unacceptable to perform abortions, so perhaps the oath actually dates to that time. (See the Appendix on page 155 for a full translation.)

OTHER OLYMPIAN HEALERS

While Hippocrates is by far the best known of the ancient Greek physicians, he was not alone in his work. There were others physicians who contributed to medical progress. The healers of note include the following:

Empedocles of Sicily (490–430 B.C.E.) developed key understandings of physiological processes of the body, including how digestion worked and the cooling function of breathing. Empedocles also developed the concept that the liver made the blood that nourished the tissues. He built on the theory of the four elements (fully explained in chapter 5's sidebar, "The Four Humors"). Legend has it that to prove his own immortality, he jumped into an active volcano, fully expecting to come back as a god.

Philosopher and medical theorist Alcmaeon of Croton (sixth century B.C.E.) wrote *Concerning Nature,* which may be the earliest example of Greek medical literature. He performed very early dissections. In his work, he located and identified the optic nerve, and he came to the conclusion that the brain controlled smelling and hearing functions. He also came to believe that the brain was the seat of sensation, not the heart as had been believed. Later, this theory fell into disfavor. He also suggested that illness resulted in internal imbalances but that they could come from external causes. Alcmaeon was very interested in anatomy and embryology, and he was one of first physician-philosophers to exert influence.

Dioscorides (40–90 C.E.) was a Greek physician who worked primarily in Rome during the time of Nero. He made great inroads into the field of pharmacology. He traveled with the military, providing medical aid as needed, and because of the vast territory of the Roman Empire, Dioscorides was able to gather and catalog a wide variety of medicinal plant substances. He also emphasized investigation and experimentation, new concepts for this time. Between 50 and 70 C.E., he wrote a fundamental five-volume study about the "preparation, properties, and testing of drugs" that became known by its Latin name, *De materia medica.* For the next 16 centuries, it was used as a central work in pharmacological studies in both Europe and the Middle East.

Herophilus of Chalcedon (335 B.C.E.–280 B.C.E.) and Erasistratus of Ceos (304 B.C.E.–250 B.C.E.)

From the date of the death of Alexander the Great (323 B.C.E.) to about 30 B.C.E., when the Romans annexed Egypt (a time known as the Hellenistic period), two physicians expanded the knowledge of how the body worked by performing dissections. Cutting into the body, even after death, was illegal until this time, so Greek knowledge had been limited by what physicians could ascertain by studying the body's exterior. Though not contemporaries, Herophilus (335–280 B.C.E.) and Erasistratus of Ceos (ca. 304–250 B.C.E.) are often linked in discussions because of the exploratory paths both chose to follow. Both Herophilus and Erasistratus were also founders of schools of belief that lasted until the second century.

Animals were most frequently used for dissection by both men, leading to misinterpretations when the information was applied to humans. Dissecting a human was still rare; however, Herophilus occasionally did public demonstrations that involved cutting up human cadavers. Erasistratus may have experimented on cutting up living animals and perhaps people (*vivisection*). Though both wrote extensively of their studies, most of their work was lost, and, therefore, their discoveries are left to the reporting of others.

Herophilus was a student of Praxagoras of Cos. Praxagoras I improved on Aristotelian anatomy by distinguishing arteries from veins, and Praxagoras's studies led him to the beliefs that arteries were conduits for circulating air throughout the body and that veins were what permitted blood circulation. Herophilus took his teacher's thought a step further. When he explored the travel of blood throughout the body, he noted that arteries carry blood, not air. He also used a water clock to study pulse rate. The water clock, which was a small earthenware vessel that measured time through the even drip of water from a hole in the bottom, was an early time-measurement system that was more accurate than a sundial in keeping track of small increments of time. Herophilus found ways to document pulse strength and rhythm, which were exciting developments for the time.

Herophilus disagreed with Aristotle's claim that the heart was the seat of intelligence, and he concluded that the brain was the center of the nervous system. He also differentiated between tendons and nerves. He examined and documented the layout of the organs, dissected an eye, and named many of the structures he discovered, including the prostate gland and the duodenum (its name is from the Greek for 12 fingers, which was the length of the organ).

Herophilus was originally a follower of Hippocrates. While he encouraged students to learn about dietetics, as Hippocrates did, he differed from the great teacher on the use of pharmaceutical cures and felt that medicines should be an available option. He also favored bleeding for cure. He is often quoted for having noted, "The best physician is the one who is able to differentiate the possible and the impossible." He authored *On Dissections* and was recognized as the father of anatomy.

Erasistratus, who was born approximately 30 years after Herophilus, was said to have been a gifted medical practitioner who believed that hygienic living led to good health. When disease occurred, Erasistratus felt it was from an accumulation of blood in a particular area or perhaps a blockage of air (*pneuma*) or fluids. Excessive phlegm caused colds in the winter; an excess of bile caused stomach distress in the summer. When he taught, Erasistratus compared blood circulation to the circulation of sap in plants. He mapped the veins and arteries that he found, and he studied the heart, eventually realizing that the heart was like a pump. The Greek philosopher Democritus (approximately 460 to 370 B.C.E.) had introduced the "atom theory," that all of mat-

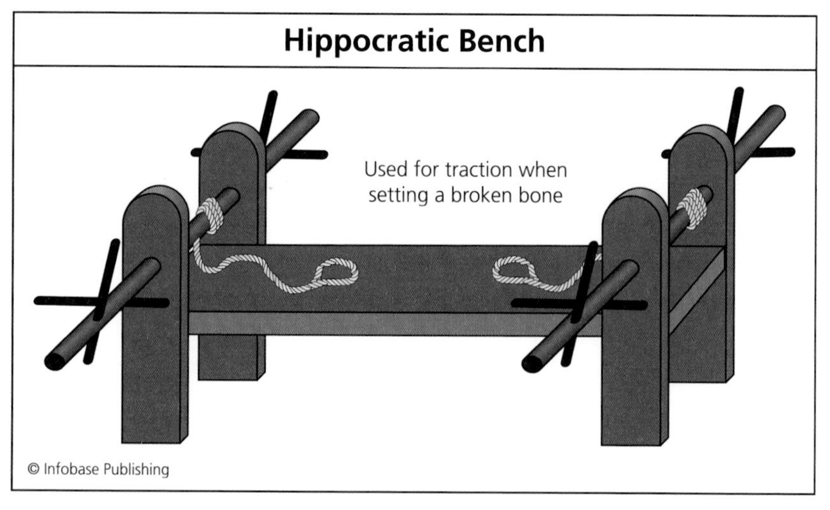

Invented by Hippocrates, the bench was an early form of traction, using tension to set bones.

ter was made up of tiny, indivisible units. Erasistratus came to believe something similar occurred with the body, that it consisted of working units that were too small to see.

Erasistratus agreed with Herophilus that the brain was the seat of intelligence, and his study of anatomy helped him realize that the sensory and motor nerves existed separately. While Erasistratus is said to have written more than 50 books on topics such as bloodletting, paralysis, drugs, poisons, and dietetics, only fragments survive. Later scholars have referred to him as the father of physiology.

Erasistratus was heavily criticized by Galen for not following Hippocrates' teachings. While Galen himself took the field of medicine forward, he did all he could to eliminate the contributions to progress made by others.

SCHOOLS OF THOUGHT DIVERGE

As the time since Hippocrates extended, physicians began to develop new ideas, and later on, medical beliefs split into different groups. Followers of Hippocrates came to be referred to as Rationalists (or Theorists). They believed in a patient-centric form of medicine and thought that disease was rooted in a scientific cause.

In approximately the third century B.C.E., the Empiricists evolved and revolted against philosophical speculation about disease as well as scientific experiments such as dissections. The Empirics felt that the best approach for a physician was to develop a treatment based on observation. Case histories became the core of their arsenal of medical tools. (Galen was later to reject their practice as "shallow.") Though their outlook was more limited, their work furthered the science of symptomalogy, and because they were open to various types of treatments, they also made progress in pharmacology and surgery. Since the Greeks had not pursued surgical cures, the Empiric efforts at performing ligatures and early operations on the goiter, mending hernias, and removing cataracts and tonsils were all instructive.

The Pneumatists was founded by Athenaieus of Attalia during the first century C.E. They regarded pneuma—air or spirit that they believed flowed through the arteries to sustain life—as the fifth element that properly needed to be added to the four humors. To Pneumatists, disease was based on whether or not the pneuma flowed freely. Archigenes (ca. 100 C.E.) was a skilled surgeon who did amputations and ligatures and was a well-respected Pneumatist.

Ultimately, what became known as the Methodical School came to dominate medicine: It was founded by Themison of Laodicea in about 50 B.C.E. The Methodists reduced the theory of medicine and therapeutics to a very few simple methods. Pulling away from the belief in the four elements, they built a practice based on part of Democritos's atom theory that taught that the cleanliness of the skin pores was the key to good health. The Roman writer Celsus (first century C.E.) felt that Methodism came about because of the increasing number of slaves and the need to treat the slaves, and this method provided an inexpensive method with the side benefit of encouraging cleanliness. The most famous Methodist was Soranus, the biographer of Hippocrates, who contributed to the fields of gynecology and obstetrics. (See chapter 8.) Some of his books still survive, though he later broke from Methodism when he realized the limitations of this school of thought.

CONCLUSION

Medicine owes a great debt to Hippocrates for introducing the art of clinical inspection and observation and stressing importance of respect for the patient. While Galen of Pergamum (introduced in the next chapter) brought forward many of the teachings of Hippocrates, the ancient world teetered between early scientific medical thinking and a continued reliance on religious healing. Desperation during the plague (see chapter 5) and later with the spread of malaria meant that the cult of Asclepius grew stronger. As Rome began to overpower the Greeks, some of the scientific medical beliefs continued, but the building of temples for the worship and healing powers of Asclepius was to continue to grow.

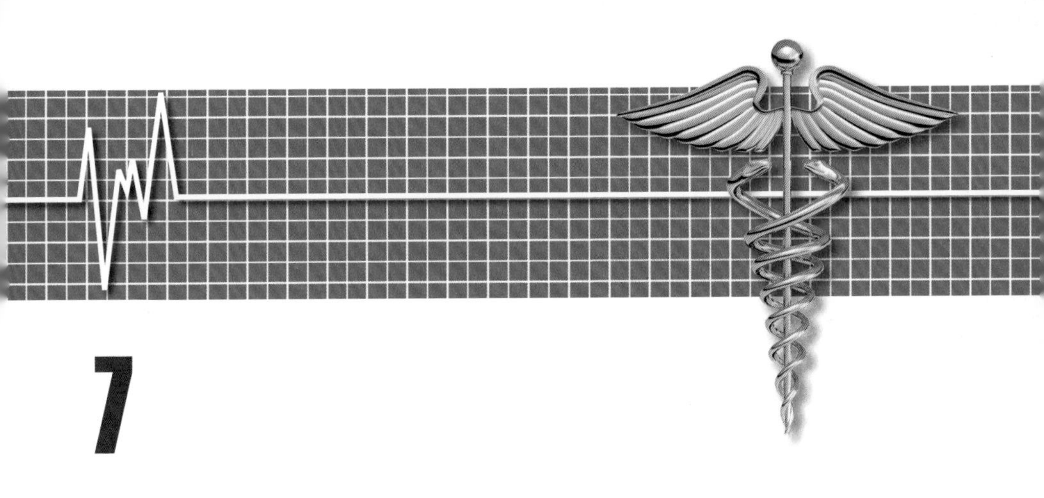

7

Galen: The Physician Whose Theories Dominated Medicine for 1,500 Years

Claudius Galen (129–199 C.E.) is one of the most influential physicians in the world of medicine. He lived 600 years after Hippocrates, but he became a strong believer in Hippocratic ideals and set about to revive many of the Greek physician's original teachings, including Hippocrates' belief in the importance of "humoural balance." He became a skilled pharmacist, a remarkable anatomist, and a leading scientist of his day.

Galen was tireless in his pursuit of truth and accuracy as he attempted to understand how the body works. He broke with the Greek tradition of disdaining surgery and practiced dissection daily. He located, compiled, and systemized the writings of Greek physicians who preceded him, and he introduced new ideas through his detailed studies of anatomy and physiology. His education in philosophy provided him with a clear eye and a sense of logic that permitted him to advance with his belief in observation-based medicine. He was so productive and such a prolific writer that it was said he employed 20 scribes to keep up with his output. He authored at least 300 known titles, 150 of which survive in whole or in part. The texts that survive are quite a collection, filling 22 volumes.

If Galen had merely been part of a continuum exploring the world of medicine, his contributions would have been quite beneficial. There was soundness to much of his anatomical exploration, and his analysis of illness showed great intelligence. However, Galen was so egotistical, self-assured, and bombastic about the correctness of his theories that it became unacceptable to disagree with him. His work so dominated the field that there was little opportunity for contributions from others. In time, Galen's work was translated into several languages so his influence was broad. For an astounding 1,500 years, his thinking dominated medicine, and medical progress was paralyzed for more than a century. A few of his *galenicals,* the term used for his pharmaceutical cures, were prescribed in Europe as late as the 19th century.

Because so little was known about science at this time, any scientific reasoning behind medicine could not be verified, so Galen's strong beliefs prevailed. This chapter explains who Galen was, how he came to dominate medicine, and why his authority held sway for so many years.

FROM SURGEON TO THE GLADIATORS TO MARCUS AURELIUS'S PHYSICIAN

Claudius Galen was a son of a wealthy architect who lived in Pergamum, Turkey, an area now known as Bergama. He is almost always referred to only as "Galen," because he never used "Claudius" on any of the works he authored.

As a youth, Galen studied Greek literature and philosophy and had no plans to practice medicine. Then his father had a dream about the Greek god of medicine, Asclepius. The elder Galen interpreted his dream to mean that his son was to undertake a medical career. Galen was sent to the city of Alexandria to study medicine under the scholars who worked there. During this period, Galen familiarized himself with all aspects of Hippocratic medicine, and he began to question many of the changes to Hippocratic beliefs that had occurred during the intervening 600 years since Hippocrates' death. As he became more knowledgeable, Galen worked hard to bring back the original teachings of the physician whom he considered the great master. He also broke with Greek tradition that eschewed cutting into the body, and he studied anatomy and surgery. He became known for being both careful in his work and meticulous in his notations about what he learned. As he studied medicine, he integrated aspects of his early education in philosophy to create the doctrine that he would teach to others. He also valued pharmaceutical remedies, and he traveled to India and Africa to learn about native plants and cures from those parts of the world.

When he returned to Pergamum in 157 C.E., he was appointed physician to the gladiators, where he cared for the wounds of those men who battled each other as part of a public entertainment.

Hippocrates and Galen *(Nina Aldin Thune)*

Galen viewed the wounds as "windows into the body," and this work increased his knowledge of human anatomy.

In 162, he moved to Rome to practice medicine. During his studies, he became strongly opposed to many of the schools or sects of medicine that had developed after Hippocrates. He publicly denounced many of them and actively debated members of the Methodism sect (see chapter 6), a group that did not believe that the four humors were the key to understanding disease. The Methodists felt that good health relied on the cleansing of pores, and Galen vociferously debated this theory. Galen's reputation grew. He lectured widely and conducted public dissections that always drew large crowds. In 166, he departed from Rome and returned to Pergamum with the announcement that he was leaving because of the "intolerable envy" of his colleagues. He actually may have left Rome to avoid contagion as a plague descended on the city at about that time.

In 168 C.E., Galen was invited to join a military campaign to northern Italy being undertaken by the then two co-emperors of Rome, Lucius Verus (130–169 C.E.) and Marcus Aurelius (161–180 C.E.). On the return trip to Rome, Verus became ill and died, and Marcus Aurelius asked Galen to return with him to become the court physician. Galen enjoyed the exalted status and imperial protection offered in this position, and he continued the job under emperors Commodus and Septimus Severus.

As a court physician, Galen followed his own instincts. A story goes that Marcus Aurelius was not feeling well, and so he summoned Galen and two other court physicians. The first two physicians conducted pulse checks, but when it was Galen's turn, he refused, saying that it was not necessary. Because a pulse check was a standard procedure and Marcus probably resented the fact that Galen was not paying as much attention to him, Marcus challenged Galen to explain why he differed in his approach. Galen replied that he did not know the emperor's normal pulse so knowing Marcus's current rate would not be helpful to him. Galen then speculated that the emperor had overeaten and suggested putting some salve on the emperor's stomach and keeping the area warm with a wool cover. Marcus Aurelius was said to comply and felt better the next day.

GALEN'S UNDERSTANDING OF MEDICINE

Galen's medical beliefs were firmly rooted in Hippocratic teachings, and his original concept of physiology—how the body worked—was a mix of ideas from philosophers including Plato and Aristotle as well as Hippocrates. Galen came to believe that the body contained three primary "operating" systems. Each was represented by a form of pneuma, an airlike substance that was considered to be essential to all of life. The bodily systems included the following:

- The brain and nerves, responsible for sensation and thought. The *pneuma physicon,* or animal spirit, dominated this system.
- The heart and arteries, responsible for life-giving energy. The *pneuma zoticon,* or vital spirit, contributed life energy.
- The liver and veins, responsible for nutrition and growth. The natural or animal spirit (*pneuma physicon*) was also present here.

When treating a patient, Galen emphasized clinical observation. He believed that careful examination and paying attention to all the symptoms would lead to an accurate diagnosis.

Galen's analytical system for assessing the balance of the four humors became part of medical teachings for years to come. Galen also integrated the philosophy that each humor connected with two of the four primary qualities of hot, cold, wet, and dry. He worked with a system of opposites to restore the balance of the four humors. Fever was treated with something cold; weak people were given difficult exercises to build up strength, and those with chest weakness were told to perform singing exercises, for example.

Galen differed from Hippocratic practice in one aspect of treatment: While Hippocrates felt that balance needed to be achieved in the body overall, Galen introduced that balance could be achieved organ by organ. This allowed physicians to develop organ-specific remedies, and the philosophy became a powerful influence on medicine.

LEARNING ANATOMY THROUGH DISSECTIONS

Galen championed that an understanding of anatomy was the foundation of medical knowledge, and he recognized that viewing the body's exterior or examining surface wounds did not permit a true understanding of how the body worked. In time, he began to conduct an increasing number of dissections. Because it was taboo to cut into human corpses, Galen conducted his work on animals, sometimes dissecting them while they were still alive, a process known as vivisection. He preferred to dissect the Barbary ape as well as pigs, both of which he thought were anatomically similar to humans. He also studied sheep and goats and eventually worked on one elephant. Mistakes were inevitable in those areas of the anatomy where animals and humans are different, but overall, Galen identified more about anatomy in general than those who had preceded him.

Unlike his contemporaries who felt surgery should be left to "journeymen," Galen felt surgery was one of the most important aspects of medicine, and he practiced dissection daily to increase his skills. He agreed with Aristotle on the importance of scientific investigation, and he experimented endlessly. He also gave very popular public demonstrations that involved cutting into living animals to demonstrate to audiences how certain parts of a body worked.

In his work, he identified seven pairs of cranial nerves, recognized that the heart worked through a system of valves, and noted that the arteries and veins differed in structure. He corrected a good number of misunderstandings in the process. For 400 years prior to Galen, medical students had been taught that the arteries

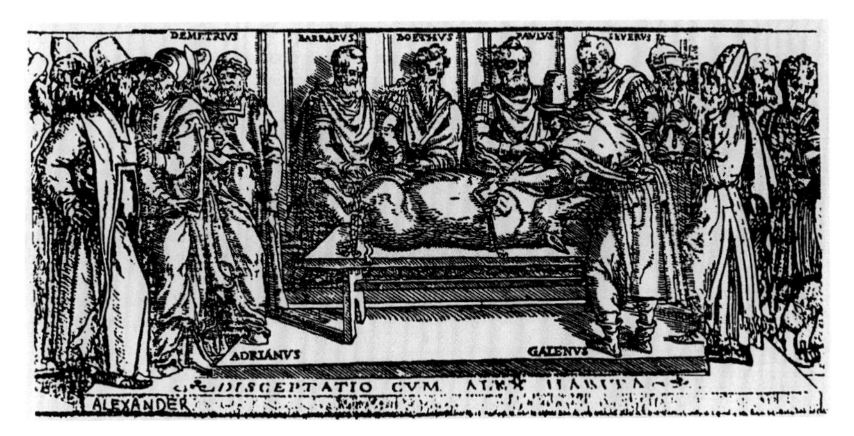

Galen pig (Image from the bottom panel of the title page to the 1541 Junta edition of Galen's Works). This depicts Galen demonstrating that the recurrent laryngeal nerves render an animal voiceless when cut.

carried air, but in his dissections, Galen proved that they carried blood. Also, since the time of Aristotle, students learned that the voice originated from the heart, but Galen proved that this was not so. One of his public demonstrations involved tying off the laryngeal nerve of a live pig. He then severed the nerves in the neck one by one while the animal continued to squeal. Finally, Galen would cut the laryngeal nerves, which resulted in an end to the squealing. A similar demonstration involved tying off the ureters in an animal—the ureters are muscular tubes that transport urine from the kidney to the bladder—to prove that urine was created by the kidneys, not the bladder. Galen also established the various functions of the spinal nerves by severing the spinal cord at various junctures on live animals, thereby proving which nerves controlled which muscles. The crowds were fascinated by these demonstrations.

He also made advances in understanding how the muscles worked, identified the structure of the skeleton, and studied the eyes, the tongue, and the reproductive organs.

Galen's accomplishments were many, but there were missteps. Galen agreed with the common misconception of the day that pus was very important to healing. Unfortunately for many injured or ailing people, wounds frequently were permitted to become infected for the pus to form. This belief persisted until the 19th century.

Galen's prestige was so great that no one questioned what he said, even when another possibility was visually obvious. Despite skeletal proof that the humerus—the upper arm bone—is straight, Galen taught that the humerus is curved. When shown examples of the bone being straight, Galen responded that those examples were "tricks of nature."

Galen made great progress with the study of the heart and the blood, but according to Galen, blood was made in the liver (See sidebar, "The Movement of Blood") and thought that the heart provided the heat of the body. He noted both the pulse and the beating of the heart but did not connect them with the movement of blood through the body. While there were inaccuracies, his studies contributed to a greater understanding of the cardiovascular system.

GALENICALS

Unlike Hippocrates, Galen believed in the power of pharmaceutical cures. He maintained a large garden to grow what he thought would be helpful in his medicines, and he worked carefully to develop new medicines. He also used and classified plants from other countries, classifying 550 species, including many sent home by Alexander the Great from India. The medicinal mixtures that he assembled generally contained a minimum of 25 ingredients and often included even more ingredients than that. Willow bark, laudanum, and opium tincture were the types of ingredients that were used popularly in that day.

Galen was meticulous in his categorization of the various medicines. He classified them by their properties—heating, cooling, drying, or moistening. He also noted their ingredients, the dosage, and how they should be administered. The galenicals, as they came to be known, were prescribed to correct

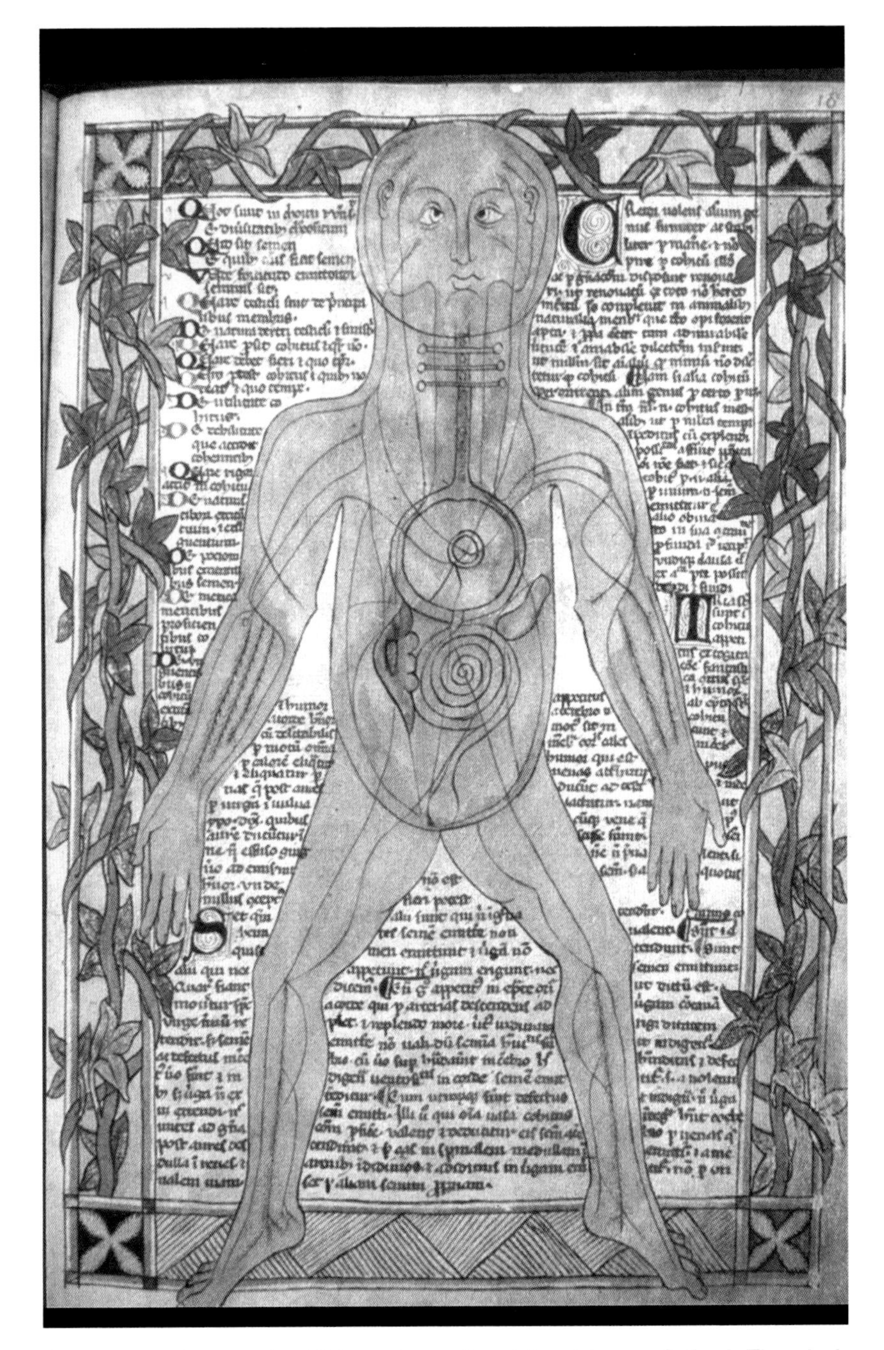

Ancient anatomy, late 13th century *(Bodleian Library, Oxford. Theodoric Borgognoni is the artist/surgeon.)*

humoral imbalances, and the philosophy of opposites was followed. An illness that was considered "hot," such as one that had a high fever, was treated with a galenical that was thought to be cooling.

THE MOVEMENT OF BLOOD

Medical practitioners who preceded Galen were seriously hampered by the fact that the study of the anatomy was limited to exterior examination of the body. As a result, there were many misunderstandings of how the body worked, including a misinterpretation of the purpose and the pathway of blood flow. Galen's information advanced understanding, but he still did not have the correct answer. In addition, he observed the pulse but did not understand how it connected with the heart.

Before Galen, there were many theories about veins and arteries and how and why blood traveled through the body. Galen believed that food that was being ingested was continuously being turned into blood. The blood then traveled via the veins to all parts of the body. Some of the blood seeped through the lungs between the pulmonary artery and pulmonary veins where it mixed with air, the *spirito vitale*, and then traveled from the right ventricle of the heart to the left through minute pores in the wall between the two chambers. From there it went through the arteries and to the brain via a special vascular network (the *rete mirabile*) situated in the neck, an idea that Galen got from dissecting hoofed animals. Along this pathway, the blood was enriched by another spirit, the animal spirit. In the brain, the blood was filtered so that impurities were discharged through a sievelike piece of bone. The impurities gave rise to tears, saliva, mucus, and sweat. Finally, the blood was thought to travel through a third vessel system, the nerves, before it reached the body's organs to provide life.

These recipes were followed through the Middle Ages, and the term *galenical* came to refer primarily to Galen's own mixtures but also to any medicine that contained a high number of ingredients. A good number of these medicines would have been some-

During its journey, the blood totally dissipated, and so, according to Galen, the body had to generate great quantities of new blood continuously. This belief contributes greatly to understanding the thinking behind bloodletting, a cure that was used both before and long after Galen. If blood was constantly being created, then the thought of releasing "excess" blood might have made sense. Galen was a big advocate of bloodletting, and this practice was followed for thousands of years.

Galen also specified that there were two types of blood: dark red venous blood and brighter and thinner arterial blood, and he noted that they each had a separate function. Venous blood was thought to originate in the liver, and arterial blood in the heart. The blood flowed from these organs to the other parts of the body where they were consumed.

Today, it is known that all blood enters the heart (the right atrium) through two veins, the superior vena cava, which gathers oxygen-poor blood from the upper half of the body, and the inferior vena cava, which gathers it from the lower half. When the right atrium contracts, the blood travels through the tricuspid valve and into the right ventricle. From the right ventricle, it is pumped through the pulmonary valve into the pulmonary artery, where it goes into the lungs to pick up oxygen. The oxygenated blood then travels back into the left atrium of the heart and on to the left ventricle before entering the aorta and traveling out to all parts of the body via the arteries.

what effective at healing or would have provided pain relief, as in the use of opium. However, many of the potions were totally without value.

Theriac was a medication that gained great popularity and was used for years as a treatment for bites and to prevent against poisoning. It was given in various forms, sometimes consumed by mouth, but other times it was mixed up to be used as a salve or plaster. Galen wrote an entire book (no longer available), *Theriaké,* extolling its use. Based on Galen's advice, Marcus Aurelius took it regularly.

Theriac, or Venice Treacle as it came to be known, took months to prepare. Depending on the preparer, theriac was assembled with at least 64 ingredients, including goat dung, pieces of mummy, and adders' heads as well as viper's flesh, opium, and a long list of special spices and herbs; cinnamon, mushrooms, fungi, and gum Arabic were among them. The high number of ingredients and the fact that many of them were rare meant that locating all the ingredients and preparing the mixture was time-consuming. Once the ingredients were gathered, they were mixed into a broth, and the broth had to be brewed and left to ferment, preferably for several years. As with other medicines that were boiled, the very act of heating the mixture for a time meant that some of the impurities were destroyed, thereby reducing some of the dangers of the ingredients. Theriac was eventually reduced to a pasty substance that could be mixed into a salve or a drink.

As a result of the complexity and time involved, theriac was very costly. It was so well regarded that in small towns in Italy, the town magistrate was placed in charge of supervising the annual brewing of it, and it was then sold in the local pharmacies. It was popular through the end of the 18th century, and as late as 1884, it was still available.

GALEN'S CONTEMPORARIES

Galen's name predominates in the history of medicine, but there were other Greeks of the time who contributed their efforts. Most of the work of Galen's contemporaries survives only in fragments, so there is little depth to what is known about them. Galen's prominent position, the popularity of his lectures, his prolific writing, and his very definite effort to belittle anything that was accomplished by anyone else served to place them even further in his shadow. Those who worked in the field during the same time included:

Rufus of Ephesus (70–120 C.E.) was a well-respected Greek physician who lived before Galen. He also believed in the study of the anatomy, accepted the doctrine of the four humors, and also cured by the use opposites. He dissected pigs and monkeys and wrote about the various body parts. He also described in detail the lens of the eye and its membrane. His writings indicated an understanding of the difference between diastolic and systolic blood pressure. He also wrote about the importance of physicians asking questions of the patient to better understand their state of health. While this is a part of modern-day medicine, physicians in ancient times were inclined to think that they could ascertain illness from physical evaluation alone. Galen praised Rufus's work, but Galen was very specific in noting that his own accomplishments overshadowed all others.

Artaeus of Cappadocia (ca. 140 C.E.) wrote in Greek and alluded to Hippocrates in his work. His book, *Acute and Chronic Diseases,* provided the best description of diseases of ancient times. Modern physicians recognize diabetes, epilepsy, tetanus, and mental disorders from his descriptions. He followed the pneumatic school, which taught that any change in the pneuma (spirit) led to illness.

Soranus practiced in Ephesus (ca. 100 C.E.) and was the chief proponent of the Methodist school of medicine. He wrote *Gynaecology,* which advanced the medical treatment of women. Works that preexisted Soranus were written with a male prejudice, and while Soranus was more compassionate in writing about women's issues, his accuracy left something to be desired. Soranus noted that babies born at seven months were more likely to live than babies born at eight months, and he described the "wandering womb" for a hysterialike illness. Soranus also invented the speculum, which eased the examination process, but after this time period, this instrument fell into disuse until late in the Middle Ages.

GALEN'S INFLUENCE

Galen wrote on almost every aspect of medicine, and his works were circulated widely during his lifetime. His followers admired his methodical and analytical techniques of identifying and curing an illness, his independent judgment, and his cautious empiricism. Because Galen was adept at promoting his own genius, his ideas traveled far, and many people became reliant on his work. By 500 C.E., his texts were heavily used in Alexandria, and his ideas soon spread to the Byzantine world. The Arab world was committed to advancing medical science, so the fact that they used Galen's teachings as their foundation for medicine prolonged the continuity of his teachings. The Arabs adapted Galen's emphasis on experimentation and empiricism, which led to new results and new observations. Starting in approximately 850 C.E., Hunayn ibn Ishaq, an Arab physician (see The Middle Ages, chapter 8) translated and annotated 129 of Galen's works. Later Rhazes, also known as Muhammad ibn Zakariya Razi, wrote "Doubts on Galen," where he notes that Galen's works should be used as a challenge for further inquiry, not accepted unquestioningly as so many were doing.

Though medical progress in western Europe stalled after the collapse of the Roman Empire, Western practitioners eventually began to observe the systemized methods being used in the Arab world. They began to take the Arabic versions of Galen and translate them into Latin; these became the basic texts for teaching medicine in the medieval universities.

In the early stages of the scientific revolution, the physician and alchemist Paracelsus (1493–1541) introduced the concept that disease resulted from agents outside the body, not from imbalance, but the belief in Galen's theories continued. In 1490, some Italian humanists turned to the original Greek writings of Galen where they began to see some of the mistakes that were made during the process of translation. Ironically, this new interest in Galen was what began to lead finally to the questioning of some of his discoveries and beliefs. When Flemish physician Andreas Vesalius devoted himself to translating Galen into Latin (ca. 1543), Vesalius began to note errors in Galen's work. Though he was greatly influenced by Galen as he wrote *De Fabrica,* he also corrected what he knew to be wrong. Galen's ideas regarding how the body worked lasted another century until the English physician William Harvey (1578–1657) identified a more accurate understanding of the flow of the blood.

Though medical thinking finally began to shift away from Galen in the 15th century, there were many aspects of his medical beliefs that survived for a very long time, including the importance of a balance of humors. As late as the U.S. Civil War, paintings of Confederate general Stonewall Jackson (1824–63) depict the general as always riding with one arm raised. It was said that he believed that this was key to maintaining humoral balance and, therefore, good health.

CONCLUSION

Galen set the medical world on a more positive path. While Hippocrates introduced medicine and healing as an art form, Galen brought forth the understanding of medicine as a science. Despite his vast overreach, some of the misinformation attributed to Galen had to do with his followers and their interpretations of his work. Had those who followed him been willing to use his thinking as stepping-stones to new ideas, then Galen would simply have been part of a continuum of medical learning. In the period following Galen, three important developments occurred in the field of medicine:

1. Medicine was increasingly taught through book learning as opposed to learning by apprenticeship.

2. The canons of Hippocrates and Galen were front and center in these studies.
3. Galen called for philosophy-doctors, a message that followers eventually interpreted as a need for physicians to study philosophy first. While this may have been a good intention, the actual process sometimes may have led to too much philosophizing and not enough attention to the person who was ill.

The misfortune of this era was the fact that Galen's influence was so strong—and remained so powerful—that medical thinking was paralyzed.

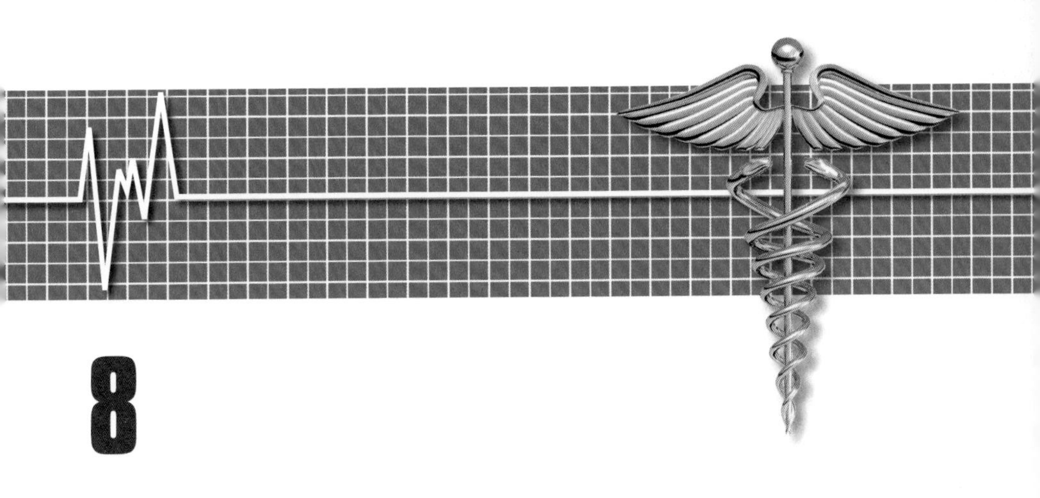

8

Medical Progress during the Roman Empire

The Roman Empire's primary contributions to health were in the field of public health. Though other civilizations had made efforts to pipe clean water into the homes and to get rid of waste properly, no one did so as successfully as did the Romans. Inspired by their Etruscan predecessors, they built *aqueducts* that delivered water to the towns, created sewage systems that removed waste from where they lived, and established bathing installations throughout the empire to encourage personal hygiene. Because so much disease in ancient times was spread by unsanitary living conditions, these advances in engineering improved the health of the people.

While the Greeks had valued art, philosophy, and science, the Romans emphasized the utilitarian, and they focused on law, government, military might, and architecture. Their accomplishments also included magnificent roadways that were built so that they could move their military into an ever-expanding area to increase the empire. This development had a direct effect on Roman health. As the armies traveled to unfamiliar parts of the world, they were exposed to new diseases. They brought back illnesses that caused high death tolls in their armies and among their communities.

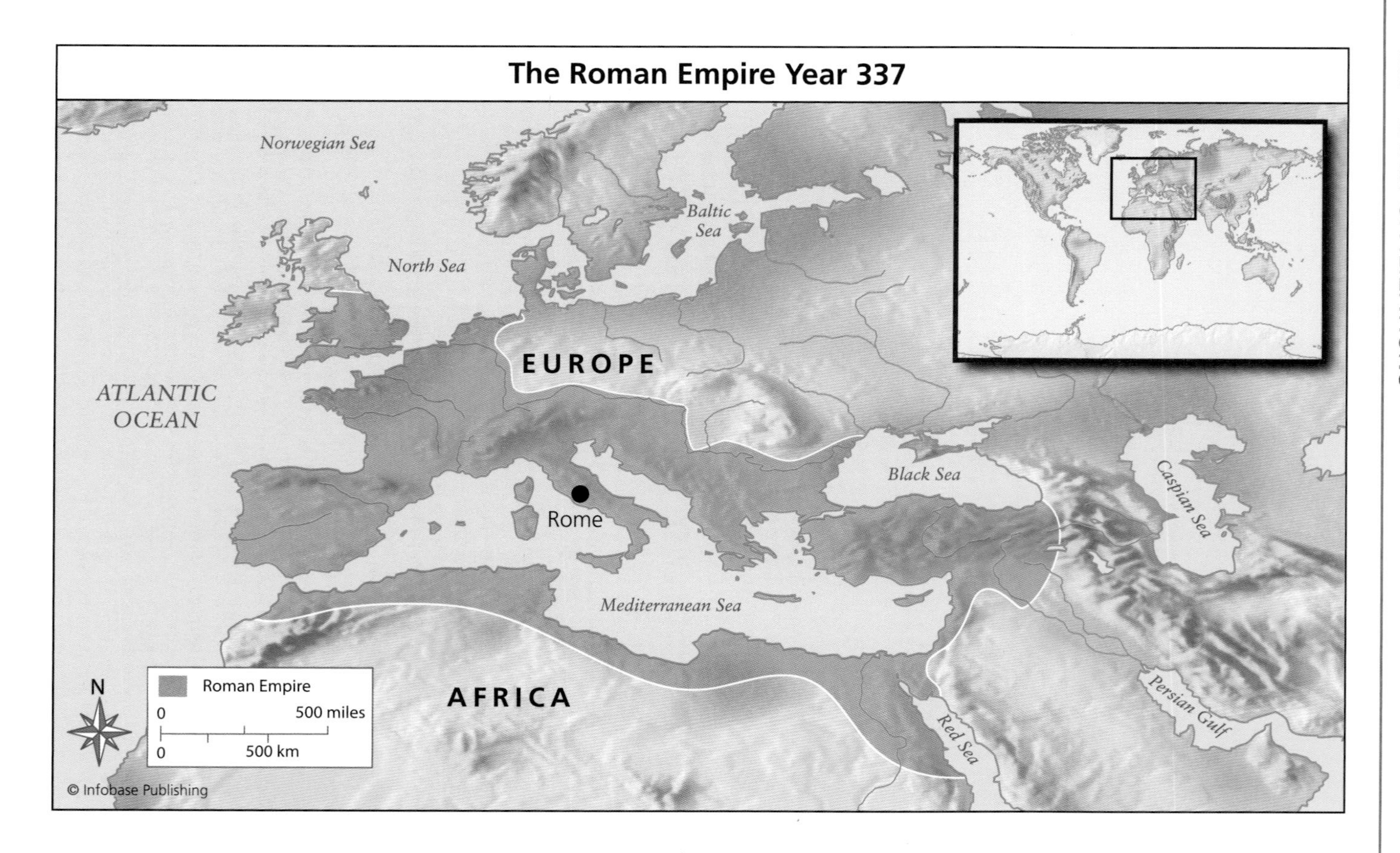
The Roman Empire Year 337
Norwegian Sea
North Sea
Baltic Sea
EUROPE
ATLANTIC OCEAN
Rome
Mediterranean Sea
Black Sea
Caspian Sea
Persian Gulf
Red Sea
AFRICA
Roman Empire
0
500 miles
0
500 km
N
© Infobase Publishing

Eventually, the Romans began to rely on their Greek predecessors for what the Greeks had learned about medicine. As the Roman Empire grew, the Romans took with them many of the Greek ideas, which they had adopted regarding medicine. A new health innovation, the hospital, was a convenience offered to the Roman military and eventually to civilians.

This chapter outlines the Roman engineering accomplishments that established a superior public-health program and discusses the Roman commitment to maintaining good health through diet and exercise. The chapter then concludes with an explanation of how the expanded Roman Empire and the relative ease of travel throughout the area eventually led to an increase in exposure to disease.

PUBLIC HEALTH AND INFRASTRUCTURE

The Romans were the first people to institute a public-health system to serve people of all income levels. They realized that to maintain good health and to be able to serve the empire, soldiers and servants needed health considerations that were similar to those of the wealthy. They built aqueducts for water delivery, created public baths to improve public cleanliness, and created methods of community waste removal.

Aqueducts

The Romans understood the importance of clean water long before they developed the aqueduct system for water delivery. Vitruvius (ca. first century B.C.E.), the Roman architect, wrote: "We must take great care in searching for springs and, in selecting them, keeping in mind the health of the people." Cities, towns, and even military forts were all built near a freshwater source. However, as

---

(Opposite) The Roman Empire helped spread medical knowledge as well as Roman sanitation methods across the area that is now Europe and the Middle East.

Pont du Gard, famous Roman aqueduct in France

these communities expanded, the dumping of wastewater and the presence of humans and animals nearby inadvertently polluted the waters. The community residents found that it was increasingly difficult to obtain freshwater.

The Romans began to explore ways to transport clean water. Underground piping was preferable as the water was then less likely to be contaminated by animal waste, and engineers eventually found methods for boring channels through rock. The channels had to be built carefully on a gentle slope so that water ran slowly but freely to its destination. Vertical shafts were bored at intervals to provide ventilation and access to the conduits. Aqueducts were built to carry the water across valleys. When the water arrived in the city, it was fed into smaller bronze or ceramic pipes where it traveled to public cisterns, the royal palaces, and the homes of the wealthy.

The conduits and aqueducts created to deliver clean water to communities were remarkable for their engineering ingenuity as well as for their massive scale. The Romans built 14 great aqueducts that carried millions of gallons of water each day—some are still used. The first aqueducts were built in 312 B.C.E., and by 1000 C.E., 250 million gallons (almost 100 million l) of water traveled through the aqueducts to the city.

As the Roman Empire expanded, engineers traveled along with the military, creating ways for the soldiers to obtain clean water and teaching local people about the merits of the water-conveyance systems. When Britain came under Roman rule, they adopted the water engineering system, creating clay-lined channels to carry the water since clay was available. The water channel supplying Dorchester was nearly eight miles (almost 13 km) long. In the town, the main pipelines were made of timber and lead and ran along the major streets with side channels that diverted water to individual buildings. The system was brilliant. If there was a drought, the channels to the individual buildings were cut off; yet

Roman aqueduct in Segovia

what water there was could still be made available to the public in town wells.

The design and building of the water system itself was an amazing feat, but the Romans also understood the importance of maintenance. They built channels with access for repairs, and engineers were assigned the job of checking for and overseeing any pipes or channels that needed to be fixed.

In 64 C.E., a major fire destroyed much of Rome (this was the fire during which ruler Nero was supposed to have fiddled). The need to rebuild the city offered the Roman people an opportunity to build their city again with straight streets and large squares as well as well-designed water-delivery and sewage-removal systems.

The new system was overseen by Julius Frontinus (ca. 35–ca. 103 C.E.), who was appointed water commissioner (*curator aquarum*) for Rome in 97 C.E. The Greek geographer Strabo (64 or 63 B.C.E.–after 23 C.E.) noted in his writings that water came in such quantity that almost every house had cisterns and water pipes and fountains.

Roman Baths

Most Roman settlements contained some sort of public bath since personal hygiene was very important in the Roman Empire. The cost of bathing was intentionally kept as low as possible to encourage the use of the baths by rich and poor. The sick were encouraged to bathe as a way to bring back their good health.

By end of fourth century C.E., there were 11 public baths and 926 private ones in Rome alone. Excavations show that a typical bathing structure contained multiple rooms, including a pool, a garden, a library, a lecture hall, an exercise room (gymnasia), a steam room, and a massage area, as well as bathing areas for hot, tepid, or cold baths. Rooms were kept warm via hot air circulat-

(Opposite) The Roman baths were quite complex structures with space for bathing, reading, exercising, and enjoying oneself.

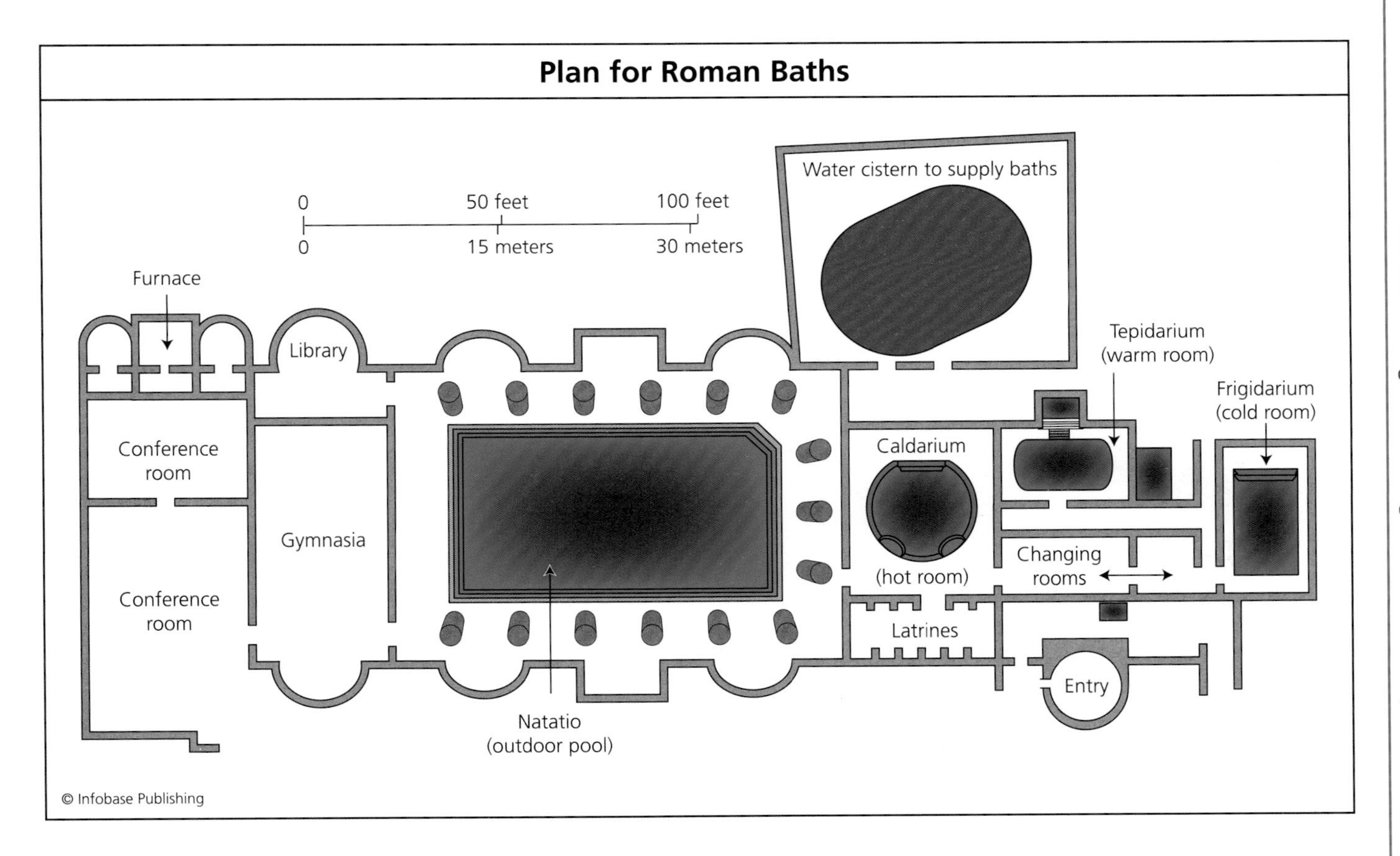
Plan for Roman Baths
0
50 feet
100 feet
0
15 meters
30 meters
Water cistern to supply baths
Furnace
Library
Tepidarium
(warm room)
Frigidarium
(cold room)
Conference
room
Caldarium
Gymnasia
Changing
rooms
(hot room)
Conference
room
Latrines
Entry
Natatio
(outdoor pool)
© Infobase Publishing

ing through pipes in walls and floors. The walls were covered with marble complemented by oversized mirrors. Even in the public baths, silver was often used for the taps for dispensing water. The private baths of the wealthy had rich decorative detail as well as amenities such as fountains and waterfalls. The largest of the Roman baths were the Baths of Diocletian, completed in 305 C.E., a lavish structure that could accommodate 3,000 bathers.

The baths were intended for relaxing and socializing as well as for cleanliness. Personal trainers and masseuses were available at some of them. Men and women had separate facilities, and children were not permitted at all.

Sanitation Engineering and Waste Removal

The Romans understood the value of removing waste products from community living areas, and they built drainage systems for this purpose. Seven rivers were made to flow through the city's sewers and flush sewage away. Most sewers had manholes that provided access so that engineers could go down into them for cleaning, maintenance, or more extensive repair work.

Initially, the engineering and design of the sewers were very much trial and error. The art of building a proper drainage system depended on fast-rushing water to wash away sewage, but in some locations the sewers were so wide that the water pooled instead of "rushed." As a result, the sewage sometimes sat in one location until there was a heavy rain. Ancient people also did not understand about microorganisms. When they channeled the wastewater to the river, people often became sick from it, but it took a long time before the stench was such that they realized that they were polluting the drinking water. An additional problem with early sewers was the rough surface of the materials that were used to create the sewer channels. The rough stone created a natural trap for microbes, so sewers sometimes became breeding grounds for disease. York, England, part of the Roman Empire, had a very high rate of cholera during the hot summers as a result of poorly made sewers.

Romans were not the first people to use toilets, however; earlier civilizations had only placed toilets in the homes of the wealthy. The Romans expanded their use by placing them in public and private buildings as well as in military forts. By 315 C.E., there were 144 public toilets in Rome, including public pay toilets at busy intersections. Within the public baths, the toilets were in a separate room built in a long row, offering facilities for as many as 20 people at one time. Other than being in a separate room from the public baths, there was no enclosure for privacy. The toilets flushed, using water that had been previously used in the baths.

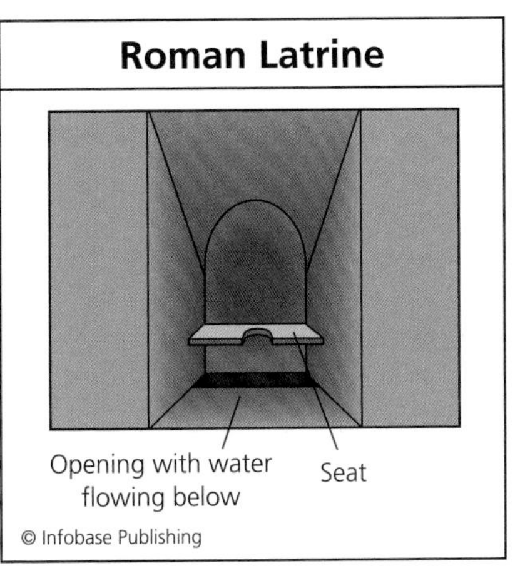

Roman latrines had good drainage systems beneath them. Some homes of the wealthy had private facilities, and the Roman baths had long rows of public latrines.

PHYSICIANS IN EARLY ROMAN SOCIETY

In the early days of the Roman Empire, medical care was the responsibility of the head of the household. He was expected to supervise medical affairs of the family, the slaves, and the animals. In actual practice, however, the women and the slaves were the ones who tended to ailing humans and animals. Each household maintained a stock of various herbs that were thought to be helpful in curing both human and beast.

Because the profession was not viewed as an honorable one, early Romans did not go into the practice of medicine. When they actually needed outside help from a person with some knowledge of healing, they summoned those who had trained in another country, usually Greece. The first Greek doctors in Rome arrived about 200 B.C.E. Many came as prisoners of war, and the wealthy then claimed them as slaves, expecting them to do housework and care for the family.

These Greeks eventually gained their freedom and set up medical practices in Rome. They became very well regarded in Roman society, and, eventually, any Roman who decided to practice medicine discovered that the only way to do so successfully was to take on a Greek name. The Roman scholar known as Pliny the Elder (23–79 C.E.) noted that Greek doctors were very popular; one particular physician attracted greater crowds in public than any of the famous actors or chariot riders in Rome.

Asclepiades (second–first century B.C.E.) was a Greek physician who did a great deal to make medicine more acceptable in Rome. He and his followers described the Hippocratic approach to medicine as a "long wait for death," and they developed new medical theories that gave rise to the Methodist sect (see chapter 6). The Methodists believed that disease was from a mechanical disturbance of the atoms that moved through the pores of the body. Asclepiades felt that patients should be treated quickly, safely, and agreeably. He recommended gentle treatments such as exercise, improved diet, cold baths, and drinking water and wine; he rejected treatments like bleeding. Later on when Galen became a presence, these theories were viewed with disfavor since Galen was so emphatic that the Hippocratic theory of the four humors was vital to good health.

The Romans were not above seeking out religious cures when the pressure was on. In 294 B.C.E., the citizens of Rome encountered a terrible plague. Herbal remedies were ineffective at stopping the death rate, and though the Romans prayed to their own god of health, Salus, they continued to fall ill. In desperation, they turned to the Greeks and asked for help from Asclepius, the Greek god of healing. When a contingency of Romans visited an Asclepius Temple in Epidaurus, it was reported that a snake was seen boarding the Roman ship. The snake is a Greek symbol of rebirth, and the Romans interpreted this as a sign that the Greek god of medicine would help Rome. The Romans built a temple for Asclepius in Rome, and shortly thereafter, the epidemic lessened its hold. As a result of this coincidental experience, Asclepius came to be worshipped in magnificent temples throughout the Roman Empire.

Despite Roman disdain for medicine, it became common practice for the military officers to take their own personal physicians along with them on military campaigns. Scribonius Largus (1–50 C.E.), a physician and pharmacologist, accompanied Claudius, emperor of Rome in 43 C.E. on his campaign to take control of Britain. From a surviving text of this time written by Scribonius, scholars have learned a great deal about this period of medicine. The work reflects Scribonius's grounding in Greek methods, expresses a strong commitment to ethics, and commits to the Hippocratic Oath. Scribonius also noted 271 remedies for many types of illness.

A 15-year civil war followed the assassination of Julius Caesar in 44 C.E., and there was a very high injury rate among soldiers on both sides. Though Roman soldiers were accustomed simply to taking care of one another, the new emperor, Augustus, formed a professional medical corps and gave the physicians dignified titles, land grants, and retirement benefits. For the next 500 years, this group helped keep soldiers on the battlefield and contributed to the advance of Roman medicine.

MEDICAL WRITINGS FROM THE ROMAN EMPIRE

After the creation of the *Corpus Hippocraticum,* there was a long lapse in medical writing until a wealthy estate owner, Celsus (ca. first century C.E.), took on the challenge of collecting and organizing the scientific works of the Alexandrian scholars. Celsus's *Artes* (*the Sciences*) was encyclopedic in scope, covering agriculture, military science, law, medicine, and philosophy. The collection originally consisted of at least 21 books, yet only the eight that were devoted to medicine survive. Without his efforts, historians would be hard-pressed to understand the medical practices of the time.

In his writings, Celsus traces medical practices from the Trojan War forward. Celsus praises the contributions of Hippocrates and the Greeks who followed him, though he makes particular note that the Roman emphasis on healthy living was important to overall health. Cures, writes Celsus, could be viewed in three categories: improving diet and lifestyle, taking medications, or undergoing surgery.

Celsus's writings contain some of the only reports of the type of surgery undertaken in that day. Much of their medical need arose from military encounters, and the intent was to get soldiers mended and back on the battlefield again. Celsus reports that the Romans used ligature for torn blood vessels and found ways to remove barbed arrows. They also could do amputations and found ways to operate on bladder stones, hernias, goiters, and cataracts and to remove venom from snakebites. The Romans used a *cupping* method to draw blood or pus from a wound.

Another important work from this era was *Historia naturalis* (*Natural history*), a 37-volume work compiled by Pliny (23–79 C.E.) that touched on many aspects of natural science, including human life and medical issues. Pliny expressed great suspicion of doctors, viewing them as incompetent, greedy, and dangerous.

The other books that were helpful to early Romans were ones on medicinal cures. Some were about the plant sources being used for drugs, and several others concerned those that were made from animal sources.

MAINTAINING GOOD HEALTH

The concept of creating an optimum environment for good health was integrated into most aspects of Roman life. One of the early issues addressed by Romans was where to situate their homes, their communities, and their forts. The Hippocratic tract, *Airs, Waters, Places,* stressed the importance of location, and the Romans believed that it was important to be at the foot of a hill and "exposed to health-giving winds."

Three Roman authors—the scholar Marcus Varro (116–27 B.C.E.), the agriculture writer Columella (first century C.E.), and the architect Vitruvius (first century B.C.E.) —put forth the hypothesis that malarial fever was produced by small animals or insects coming out of the swamps. Varro wrote that houses and forts should not be built near swamps "because certain tiny creatures which cannot be seen by the eyes breed there. These float through the air and enter the body by the mouth and nose and cause serious disease." Columella noted similarly: "There should be no marshes near buildings, for marshes give off poisonous vapors during the hot period of the summer. At this time they give birth to animals with mischief-making stings which fly at us in thick swarms." Julius Caesar (11–44 B.C.E.) arranged for the draining of the Codetan Swamp near Rome and planted a forest in its place. The incidence of malaria is said to have been reduced from his effort.

The Romans also deduced that if the soldiers stayed too long in one place, they would suffer the illnesses of the area, so military settlements were moved regularly.

One other safety practice became a part of Roman life. Burials within the city were forbidden, and cremation was practiced, all of which created a healthier way of disposing of the bodies for that time.

Treatments

"A person should put aside some part of the day for the care of his body. He should always make sure that he gets enough exercise especially before a meal," Celsus wrote, voicing the Roman philosophy that to maintain good health, it was important to take care of the body. Certain foods were also used to fend off illness. The Roman politician Cato (234–149 B.C.E.) believed that cabbage was the key to good health, and he lived to be 84, exceptionally old for that time.

When a Roman became ill, the first line of approach was a home remedy. Many of these were used on both people and animals. In addition to specifying the ingredients, the recipe might

Roman Cupping Vessel

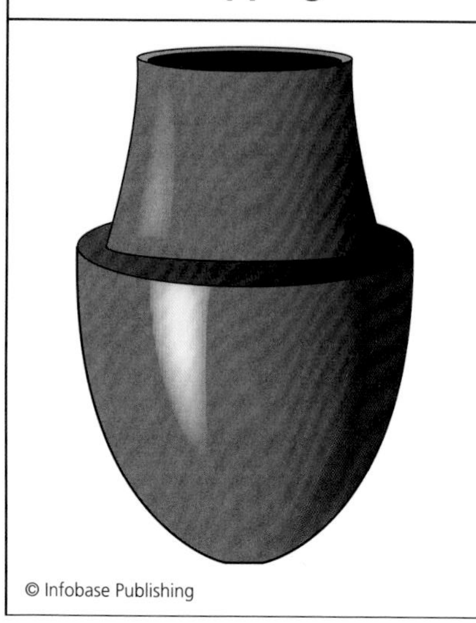

When heated, this style of vessel created a partial vacuum that would stick to the body and draw blood or pus from a wound.

note that the potion needed to be mixed by a person who was standing up, for example, or that there would be a repetition of numbers in the items used, adding an element of magic to the cure.

There were a few very popular remedies that were used commonly. For bruises and swellings, it was recommended that unwashed wool be dipped in a mixture of "pounded rue" and fat. For uterine inflammation, rams' wool that was washed in cold water and soaked in oil was thought to be soothing. Wool dipped into a mixture of oil, sulphur, vinegar pitch, and soda cured lumbago. The Romans also picked up on the Etruscan belief in hepatoscopy, which involved reading divine signals from the liver of a sacrificed animal, so on occasion, this method was used to decide on a treatment.

Ancient people understood the risks of surgery, so surgeons attempted other healing methods first. They set broken bones, and they could put back dislocated joints. Open wounds were washed with wine or vinegar to reduce the chance of infection. If they did undertake surgery, the process was a difficult one as they had no effective anesthetic, though wine or drugs such as hemlock were commonly used to attempt to ease a patient's pain. Surgical tools used on the wealthy were usually intricately carved, but, ironically, the well-to-do were more likely to contract infections from surgery because the intricate carvings held dirt and were harder to clean.

THE ROMAN VALETUDINARIAN (HOSPITAL)

Hospitals, known as *valetudinarians,* were permanent, well-designed structures originally built for the military. Some had central heating, and all were designed with good ventilation. A main hallway led to a dining hall, latrines, and baths as well as many small rooms for patient care. While they did not understand about germs and contagion, they realized that being around a person who was ill sometimes made others sick, and the separate rooms provided the opportunity to reduce exposure.

The first civilian public hospital in Rome was founded by a wealthy Roman matron, Fabiola, who converted to Christianity and decided to help the sick as penance for marrying a second husband before her first was dead. In 390 C.E., she sold her possessions to acquire money for the hospital. (A recent excavation revealed that the hospital had five wards and would have had space for at least 400 beds.) She established a convalescent home at her country villa for discharged patients who still needed care. In her effort to be forgiven for her sins, Fabiola herself tended to patients and was said to walk the streets of Rome, looking for the sick, the abandoned, and the dying to bring them in for care and comfort.

THE SPREAD OF MAJOR DISEASE

As the Romans expanded their territory, they encountered—and took with them—the pathogens of various diseases. People in one geographic area tend to build up resistance to specific illnesses. As trade and travel increased, the Romans were exposed to new pathogens, and the results often were deadly. The Antonine plague (thought to be smallpox) was one of the first serious outbreaks of epidemic disease. Between 165 and 180 C.E., it killed a quarter of the population in areas where it spread, some 5 million people in all. Another outbreak was introduced into Rome 60 years later and killed about 5,000 people per day.

Malaria was also deadly. A particularly serious outbreak occurred in the first century B.C.E., causing a fall in the birth rate that weakened the city and led to a long decline.

The first known outbreak of the bubonic plague occurred in the Roman Empire, which lost about one-quarter of its population to the disease. Though it was not understood at the time, the infection was carried by flea-infested rats that traveled throughout ancient cities. When the rats died of the disease, the fleas looked for other hosts. When they chose to travel on people, the people became infected. The disease itself became one that could spread from person to person through the air but also in clothing, making it very difficult to wipe out.

The increasing popularity of Christianity was partially due to the various plagues. Illness was seen as a consequence of sin, and people often turned to religion and prayer for healing.

GYNECOLOGY

Pregnancy, childbirth, and early infancy were very dangerous times. Mothers frequently died in childbirth, and about 25 percent of babies born in Rome died before the age of one.

During childbirth, if help was needed by a woman, midwives were responsible for overseeing the birth. While they had experience, they lacked technical knowledge, and there was a lot of misinformation. One belief was that if there were problems during labor, four people should surround the pregnant woman and shake her violently 10 times to speed the birth. Pliny the Elder wrote that to assure a safe birth, at the first sign of labor, a Roman needed to take a spearlike implement and hurl it over the house three times.

A popular myth states that Julius Caesar was born by caesarean section. This is inaccurate. While live infants sometimes were removed from the womb successfully, the surgery was only undertaken after the death of the mother. The term *cesarean section* refers to Caesarean law, which demanded that when a pregnant woman died, her body could not be

Not Just Disease Was the Enemy

Poisons were a serious concern to people of all social classes. Poor farmers understood that there were many unknowns that could harm them, and the ruling class worried about intentional poisonings. As a result, many healers focused on antidotes to poisons as well as treatments against the bites of dogs, apes, snakes, wild animals, and even human beings.

Mithradates Euphator (120–63 B.C.E.), an early enemy of the Romans, became a ruler when he was just a boy because his father had died. It was believed that his mother had poisoned her husband so that she could rule until her son, Mithradates, was older. The story goes that because of this fear, Mithradates worked to

buried until the baby had been removed. The law stated that if the woman was alive, then this type of surgery could not be performed until the mother was in the 10th month of gestation, meaning that she was in trouble anyway.

The womb was a greatly misunderstood organ. In the sixth century B.C.E. it was believed that it moved throughout a woman's body and was sometimes called the "wandering womb." The word *hysteria,* is derived from the Greek word *hystera* for womb. Aretaeus of Cappodocia, a contemporary of Galen, wrote of the wandering womb and described how it moved throughout a woman's body. He noted that these movements of the womb caused hysteria.

During the second century C.E., Soranus of Ephesus wrote about childbirth, infant care, and diseases of women. He opposed abortion and suggested numerous means of contraception. He also wrote about how to turn an infant in the womb to facilitate an easier birth. This method was to be used during his day and then was forgotten until the 16th century.

build up resistance to various poisons by taking small samples of many different things. When he and his soldiers were trapped by the Roman army, he poisoned all of his men so that they need not die at the hands of the Romans. Only he was left to be captured because he himself was immune. He eventually escaped to the Crimea and found another way to commit suicide.

CONCLUSION

With the fall of the Roman Empire in 476 C.E., Western civilization entered a very bleak period. The aqueducts and sewer systems that helped improve public health in the empire were maintained only half heartedly in most areas, and many fell into disrepair. During the Middle Ages, people's access to clean water and methods for removing waste diminished. The Greek medical beliefs that had been adopted by Romans were largely replaced by a reversion to religious healing; therefore, many of the gains made during previous years were to lie dormant for a thousand years.

APPENDIX

THE TRANSLATED TEXT OF THE HIPPOCRATIC OATH

The Hippocratic Oath is part of the *Corpus Hippocraticum,* an anthology of medical writings by several authors, some of whom were students of the Greek physician Hippocrates. Though experts do not think that Hippocrates himself wrote the oath, it is still held sacred by physicians. It is one of the oldest binding documents in history. Though there are many versions, the basic tenets remain the same: to treat the ill to the best of one's ability, to preserve a patient's privacy, and to teach the secrets of medicine to the next generation. The following is the Hippocratic Oath as it appears in one commonly used translation:

> I swear by Apollo the physician, and Aesculapius, and Health, and All-heal, and all the gods and goddesses, that, according to my ability and judgment, I will keep this Oath and this stipulation—to reckon him who taught me this Art equally dear to me as my parents, to share my substance with him, and relieve his necessities if required; to look upon his offspring in the same footing as my own brothers, and to teach them this art, if they shall wish to learn it, without fee or stipulation; and that by precept, lecture, and every other mode of instruction, I will impart a knowledge of the Art to my own sons, and those of my teachers, and to disciples bound by a stipulation and oath according to the law of medicine, but to none others. I will follow that system of regimen which, according to my ability and judgment, I consider for the benefit of my patients, and abstain from whatever is deleterious and mischievous. I will give no deadly medicine to any one if asked, nor suggest any such counsel; and in like manner I will not give to a woman a pessary to produce abortion. With purity and with holiness I will pass my life and practice my Art. I will not cut persons laboring under the stone, but will leave this to be done by

men who are practitioners of this work. Into whatever houses I enter, I will go into them for the benefit of the sick, and will abstain from every voluntary act of mischief and corruption; and, further from the seduction of females or males, of freemen and slaves. Whatever, in connection with my professional practice or not, in connection with it, I see or hear, in the life of men, which ought not to be spoken of abroad, I will not divulge, as reckoning that all such should be kept secret. While I continue to keep this Oath unviolated, may it be granted to me to enjoy life and the practice of the art, respected by all men, in all times! But should I trespass and violate this Oath, may the reverse be my lot!

CHRONOLOGY

3100 B.C.E.	Egyptians begin to bury kings and nobility in tombs.
3000 B.C.E.	First signs of smallpox and pox-type illnesses
ca. 2900 B.C.E.	Fu-His develops concept of yang and yin.
ca. 2700 B.C.E.	Shen Nong, early Chinese pharmacologist, invents acupuncture.
ca. 2600 B.C.E.	Huangdi (Huang-ti) authors *Huangdi Neijing* (*Huang-ti Neiching; The Inner Canon of the Yellow Emperor*), which is to become the classic book on internal diseases.
2613–2494 B.C.E.	Egyptians create a successful mummification method.
2400 B.C.E.	Lady Peseshet thought to be the first female doctor
1792–1750 B.C.E.	Hammurabi, ruler of Babylon, establishes rules for practicing medicine.
1825 B.C.E.	Kahun Gynecological Papyrus. These writings concerned pregnancy, birth, and female health issues.
1600 B.C.E.	Smith Papyrus, referred to as "Book of Wounds." This papyrus primarily covered trauma surgery.
1550 B.C.E.	Ebers Papyrus—remedies to get rid of demons—outlines four-step method for conducting a clinical exam.
800 B.C.E.	Homer writes *The Odyssey.*
ca. 563–ca. 483 B.C.E.	Buddhism, introduced by Siddhartha Gautama, preaches universal love, service, the banishment of suffering, and the gain of peace of mind.

sixth century B.C.E.	Alcmaeon of Croton concludes that the brain controls smelling and hearing functions and is the seat of sensation.
fifth century B.C.E.	Aslcepius viewed as god of healing; teaches the Greek philosophy: "First the word, then the herb, lastly the knife."
490–430 B.C.E.	Empedocles of Agrigetum introduces the four bodily humors: blood, yellow bile, black bile, and phlegm.
470–399 B.C.E.	Socrates develops orderly series of questions for any type of medical problem.
460–377 B.C.E.	Hippocrates of Cos teaches that diseases come from natural causes and also the importance of observing patients and their symptoms.
390 B.C.E.	First civilian public hospital is founded by Roman matron Fabiola.
384–322 B.C.E.	Aristotle teaches about the scientific method, systematic observation, and experimentation; he advances the creation of the scientific method.
335–280 B.C.E.	Herophilus of Chalcedon distinguishes arteries from veins, finds way to document pulse strength and rhythm, and concludes that the brain is the center of the nervous system.
312 B.C.E.	First aqueducts in the Roman Empire
304–250 B.C.E.	Erasistratus of Ceos discovers that the heart is a pump; he becomes known as the father of physiology.
50 B.C.E.	Methodist School founded by Themison of Tralles on the theory that cleanliness of the skin pores was the key to good health.
first century C.E.	Celsus collects and organizes the scientific works of the Alexandrian scholars.
40–90 C.E.	Dioscorides writes *De materia medica,* a study about the preparation, properties, and testing of drugs.

23–79 C.E.	Pliny the Elder writes *Naturalis Historia* about natural science, including human life and medical issues.
70–120 C.E.	Writings of Rufus of Ephesus describe the difference between diastolic and systolic blood pressure.
97 C.E.	Julius Frontinus is appointed Water Commissioner (*curator aquarum*) for Rome.
ca. 100 C.E.	Soranus writes *Gynecology,* advancing medical treatment of women.
110–207 C.E.	Hua Duo invents method of rendering a patient temporarily unconscious.
129–199 C.E.	Claudius Galen—skilled pharmacist, anatomist, and scientist—strongly supports Hippocratic teaching, clinical observation, and the importance of surgery.
ca. 140 C.E.	Artaeus of Cappadocia writes *Acute and Chronic Diseases,* describing diseases of ancient times and teaches that any change in the pneuma (spirit) can lead to illness.
150–219 C.E.	Zhang Zhongjing writes *Shang Han Lun* or *Treatise on Febrile Diseases* covering acupuncture, moxibustion, and herbal remedies.
ca. 180–270 C.E.	Wang Shuhe—authority on the pulse
283?–343 C.E.	Ge Hong, an alchemist, wrote *Zhou Hou Fang,* or the *Handbook of Prescriptions for Emergency Treatment,* a volume about the transmission of infectious diseases.
315 C.E.	Public pay toilets are installed in Rome.
476 C.E.	Fall of the Roman Empire
1493–1541	Paracelsus introduces concept that disease resulted from agents outside the body, not from imbalance.
1578–1657	William Harvey develops theory of the circulation of blood.

1859–1917	Sir Marc Armand Ruffer—first modern paleopathologist
1947	Radioactive carbon 14 is used as a method for dating artifacts and human remains.

GLOSSARY

abortion the termination of a pregnancy after, accompanied by, resulting in, or closely followed by the death of the embryo or fetus

acupressure the application of pressure (as with the thumbs or fingertips) to the same discrete points on the body stimulated in **acupuncture** that is used for its therapeutic effects (as in relief of tension or pain)

acupuncture an original Chinese practice of inserting fine needles through the skin at specific points especially to cure disease or relieve pain (as in **surgery**)

alchemy a medical chemical science and speculative **philosophy** aiming to achieve the transmutation of the base metals into gold, the discovery of a universal cure for disease, and the discovery of a means of indefinitely prolonging life

anatomy the act of separating the parts of the organism to ascertain their position, relations, structure and function

aqueduct a conduit for water; especially one for carrying a large quantity of flowing water

artery any of the tubular branching muscular- and elastic-walled vessels that carry blood from the heart through the body

auricular understood or recognized by the sense of hearing

bloodletting phlebotomy—the letting of blood for transfusion, diagnosis, or experiment, and especially formerly in the treatment of disease

caduceus the symbolic staff of a herald; *specif:* a representation of a staff with two entwined snakes and two wings at the top; an insignia bearing a caduceus and symbolizing a physician

cautery the act or effect of cauterizing; an agent (as a hot iron or caustic) used to burn, sear, or destroy tissue

charlatan one making showy pretenses of knowledge or ability: fraud, faker.

chi* or *qi vital energy that is held to animate the body internally and is of central importance in some Eastern systems of medical treatment (as acupuncture) and of exercise or self-defense (as tai chi)

cosmology a branch of metaphysics that deals with the nature of the universe

cupping an operation of drawing blood to the surface of the body by use of a glass vessel evacuated by heat

curator aquarum water commissioner

dissection an anatomical specimen prepared by dissecting

dura mater fibrous membrane forming the outer envelope of the brain

empirical originating in or based on observation or experience

epidemiology a branch of medical science that deals with the incidence, distribution, and control of disease in a population

exudates exude matter

galenicals the term used for Galen's pharmaceutical cures

Hippocratic Oath an oath embodying a code of medical ethics usually taken by those about to begin medical practice

materia medica substances used in the compositions of medical remedies; drugs, medicine

meridian any of the pathways along which the body's vital energy flows according to the theory behind **acupuncture**

midwife a person who assists women in childbirth

moxibustion the therapeutic use of moxa, a soft wooly mass prepared from the ground young leaves of Eurasian artemisia (especially *Artemisia vulgaris*), that is used in Chinese and Japanese medicine typically in the form of sticks or cones that are ignited and placed on or close to the skin or used to heat **acupuncture** needles

paleopathology a branch of pathology concerned with ancient diseases as evidenced especially in fossils or other remains

philosophy a discipline comprising its core logic, aesthetics, ethics, metaphysics, and epistemology

physiology a branch of biology that deals with the functions and activities of life or of living matter (as organs, tissues, or cells) and of the physical and chemical phenomena involved—compare **anatomy**

placebo a usually pharmacologically inert preparation prescribed more for the mental relief of the patient than for its actual effect on a disorder

plague an epidemic disease causing a high rate of mortality

pneuma soul, spirit

psychological of or relating to psychology, mental

pulse the regular expansion of an **artery** caused by the ejection of blood into the arterial system by the contractions of the heart

rete mirabile A vascular network interrupting the continuity of an **artery** or **vein**

scarification the act or process of scarifying; to make scratches, small cuts in, as the skin (for example, an area for vaccination)

scientific method a method of research in which a problem is identified, relevant data are gathered, a hypothesis is formulated from these data, and the hypothesis is empirically tested

somatic of, or relating to, or affecting the body, especially as distinguished from the germplasm or the psyche

surgery a branch of medicine concerned with diseases and conditions requiring or amenable to operative or manual procedures

symptom subjective evidence of disease or physical disturbance; *broadly:* something that indicates the presence of bodily disorder

trephination the act or instance of perforating the skull with a medical instrument; also treparíation

vein any of the tubular branching vessels that carry blood from the capillaries toward the heart

vivisection the cutting of or operation on a living animal usually for physiological or pathological investigation

wu xing a classification system used by the Chinese that involved identifying everything according to a system of five elements

yang the masculine active principle in nature that in Chinese **cosmology** is exhibited in light, heat, or dryness and that combines with **yin** to produce all that comes to be

yin the feminine passive principle in nature that in Chinese **cosmology** is exhibited in darkness, cold, or wetness and that combines with **yang** to produce all that comes to be

yoga a Hindu theistic **philosophy** teaching the suppression of all activity of body, mind, and will in order that the self may realize its distinction from them and attain liberation

ABOUT SCIENCE AND HISTORY

Diamond, Jared. *Collapse: How Societies Choose to Fail or Succeed.* New York: Viking Penguin, 2006. Diamond examines a wide range of ancient and more recent societies to determine what led to each civilization's collapse. Environmental destruction and the spread of illness are often contributing factors.

———. *Guns, Germs, and Steel: The Fates of Human Societies.* New York: W. W. Norton, 1999. Diamond places in context the development of human society, which is vital to understanding the development of medicine.

Hazen, Robert M., and James Trefil. *Science Matters: Achieving Scientific Literacy.* New York: Doubleday, 1991. A clear and readable overview of scientific principles and how they apply in today's world, which includes the world of medicine.

Internet History of Science Sourcebook. Available online. URL: http://www.fordham.edu/halsall/science/sciencsbook.html. Accessed July 9, 2008. A rich resource of links related to every era of science history, broken down by disciplines, and exploring philosophical and ethical issues relevant to science and science history.

Lindberg, David C. *The Beginnings of Western Science, Second Edition.* Chicago: University of Chicago Press, 2007. A helpful explanation of the beginning of science and scientific thought. Though the emphasis is on science in general, there is a chapter on Greek and Roman medicine as well as medicine in medieval times.

Roberts, J. M. *A Short History of the World.* Oxford: Oxford University Press, 1993. This helps place medical developments in context with world events.

Silver, Brian L. *The Ascent of Science.* New York: Oxford University Press, 1998. A sweeping overview of the history of science from the Renaissance to the present.

Spangenburg, Ray, and Diane Kit Moser. *The Birth of Science: Ancient Times to 1699.* New York: Facts On File, 2004. A highly readable book with key chapters on some of the most significant developments in medicine.

ABOUT THE HISTORY OF MEDICINE

Ackerknecht, Erwin H., M.D. *A Short History of Medicine, Revised Edition.* Baltimore: Johns Hopkins University, 1968. While there have been many new discoveries since Ackerknecht last updated this book, his contributions are still important as it helps the modern researcher better understand when certain discoveries were made and how viewpoints have changed over time.

Clendening, Logan, ed. *Source Book of Medical History.* New York: Dover Publications, 1942. Clendening has collected excerpts from medical writings from as early as the time of the Egyptian papyri, making this a very valuable reference work.

Darvill, Timothy. *Stonehenge: The Biography of a Landscape.* London: Tempus, 2007. Understanding of Stonehenge has changed over time, and Darvill provides a history of this historic site and reports on his own recent archaeological work.

Davies, Gill, ed. *Timetables of Medicine.* New York: Black Dog & Leventhal, 2000. An easy-to-assess chart/time line of medicine with overviews of each period and sidebars on key people and developments in medicine.

Dawson, Ian. *The History of Medicine: Greek and Roman Medicine.* New York: Enchanted Lion Books, 2005. A heavily illustrated short book to introduce young people to the Greek and Roman approach to medicine. Dawson is British, so there is additional detail about the development of medicine in the British Isles.

Duffin, Jacalyn. *History of Medicine.* Toronto, Canada: University of Toronto Press, 1999. Though the book is written by only one author (a professor), each chapter focuses on the history of a single aspect of medicine, such as surgery or pharmacology. It is a helpful reference book.

Internet Ancient History Sourcebook: Hellenistic World. Available online. URL: www.fordham.edu/halsall/ancient/asbook08.html. Accessed July 10, 2008. A rich resource of links related to ancient history, broken down by disciplines, and exploring philosophical and ethical issues relevant to science and science history.

Kennedy, Michael T., M.D., FACS. *A Brief History of Disease, Science, and Medicine.* Mission Viejo, Calif.: Asklepiad Press, 2004. Michael Kennedy was a vascular surgeon and now teaches first and second year medical students an introduction to clinical medicine course at the University of Southern California. The book started as a series of his lectures, but he has woven the material together to offer a cohesive overview of medicine.

Loudon, Irvine, ed. *Western Medicine: An Illustrated History.* Oxford: Oxford University Press, 1997. A variety of experts contribute chapters to this book that covers medicine from Hippocrates through the 20th century.

Magner, Lois N. *A History of Medicine.* Boca Raton, Fla.: Taylor & Francis Group, 2005. An excellent overview of the world of medicine from paleopathology to microbiology.

Porter, Roy, ed. *The Cambridge Illustrated History of Medicine.* Cambridge, Mass.: Cambridge University Press, 2001. In essays written by experts in the field, this illustrated history traces the evolution of medicine from the contributions made by early Greek physicians through the Renaissance, Scientific Revolution, and 19th and 20th centuries up to current advances. Sidebars cover parallel social or political events and certain diseases.

Porter, Roy. *The Greatest Benefit to Mankind: A Medical History of Humanity.* New York: W. W. Norton, 1997. During his lifetime, Porter wrote a great amount about the history of medicine, and this book is a valuable and readable detailed description of the history of medicine.

Rosen, George. *A History of Public Health, Expanded Edition.* Baltimore: Johns Hopkins University Press, 1993. While serious public-health programs did not get under way until the 19th century, Rosen begins with some of the successes and failures of much earlier times.

Simmons, John Galbraith. *Doctors & Discoveries.* Boston: Houghton Mifflin, 2002. This book focuses on the personalities behind the discoveries and adds a human dimension to the history of medicine.

United States National Library, National Institutes of Health. Available online. URL: http://www.nlm.nih.gov/hmd/. Accessed July 10, 2008. A reliable resource for online information pertaining to the history of medicine.

INDEX

Note: Page numbers in *italic* refer to illustrations; *m* indicates a map; *t* indicates a table.

C

D

H